TERMS OF DISSERVICE

Praise for *Terms of Disservice*

"In a superbly crafted book, Dipayan Ghosh lays out the rise—and the perils—of big tech. To the lay reader, *Terms of Disservice* provides an accessible insight into how the rise of consumer internet has upended the world as we knew it, empowered but also undermined the fundamental rights of citizens, and expanded but also threatened democracy. To policymakers, it provides a robust proposal for a new social contract which addresses the dangers emanating from the business model of select big tech firms. And to the world at large, Ghosh shows how ethical imperatives, technological leaps, economic progress and democratic functioning can and must coexist. And if it does not, the idea that citizens must, through their free will, choose governments, in the backdrop of an honest public sphere, will be further jeopardized. This is a must-read for those engaged in the world of tech policy, but beyond that, for anyone who wants to understand how new structures of corporate power are shaping the world of politics, economy, and everyday life—often with terrifying consequences. And there can be few better chroniclers of this shift than Ghosh who has straddled, with great distinction, the worlds of tech, academia and policymaking."

—Prashant Jha, Editor, Views, *Hindustan Times*

"It is now beyond doubt that the consumer internet ecosystem—driven by capitalism and largely unchecked by governments—has created a breeding ground for misinformation and the exploitation of our data. In *Terms of Disservice*, Dipayan Ghosh not only reveals the hidden motivations of major global tech firms but also lays out a detailed framework for how to regulate these behemoths in the future. This is an essential book for policymakers and general readers around the world."

—Ravi Agrawal, Editor-in-Chief of *Foreign Policy* and author of *India Connected: How the Smartphone Is Transforming the World's Largest Democracy*

"*Terms of Disservice* is a frank indictment of big tech in weakening democratic societies. Ghosh's diagnosis, and treatment, of the ills provide a credible roadmap for Indian regulators to step in where the U.S. regulators have failed to. With lax data laws in the country, it's a necessary read to understand our immediate future."

—Neyaz Farooquee, BBC World Service, and author of *An Ordinary Man's Guide to Radicalism: Growing Up Muslim in India*

"Powerful internet platforms, driven by a ruthless commercial logic, have caused tremendous harm in the United States and around the world. With technological sophistication and crystal-clear writing, *Terms of Disservice* provides the analytical tools necessary to understand the structural roots of this crisis and the policies we need to confront it. Anyone who thinks democracy is worth saving should read this compelling book."

—Victor Pickard, Professor of Media Policy, University of Pennsylvania

"Ghosh's deep knowledge of the way digital companies undermine our cognition, agency, and democracy is scary but required reading. Thankfully, his deep faith in the ability of government and civil society to develop a new social contract makes this a hopeful and actionable proposal."

—Douglas Rushkoff, Professor of Media Theory, Queens College, and author of *Team Human*

"*Terms of Disservice* addresses the big picture of social media—both its place in our national economy and its corrosive effect on the individual. Ghosh concludes with a proposal for a social contract that preserves the strengths of social media while respecting the autonomy of the individual. The result is a roadmap for policymakers—we can only hope that they pay attention."

—Stephen Wicker, Professor of Electrical and Computer Engineering, Cornell University, and author of *Cellular Convergence and the Death of Privacy*

"Dipayan Ghosh is one of the most important thinkers of his generation. His experience at the White House, in academia and in industry laid deep intellectual roots that cut across disciplines. Ghosh understands the effects of technology in a way that few do and offers a compelling vision for how to maximize its benefits and minimize its harms."

—Alec Ross, author of *Industries of the Future* and former Senior Advisor, U.S. Department of State

"With subject expertise and lively writing, Dipayan Ghosh provides a rare and important examination of the underlying problems of Big Tech today: the business models."

—Cecilia Kang, *The New York Times*, and author of *An Ugly Truth: Inside Facebook's Battle for Domination*

"*Terms of Disservice* follows the money itself in part by re-conceiving the 'currency' of this marketplace—not just as cash transactions, but as constant, frictionless, ever increasing payment sucked up by platform firms in the form of users' engagement, attention, and data."

—Andy Fitch, *Los Angeles Review of Books*

"One of the greatest challenges to confronting misinformation and disinformation is being able to define the problem in the first place—and Ghosh helps us to follow the money. We won't begin to solve this urgent problem unless we confront the market incentives that make it not just viable, but profitable. In this book, Ghosh helps us take a big leap forward in understanding the business model that underpins fake news—and how we can start to do something about it."

—Robby Mook, CNN political commentator, and Campaign Manager of Hillary Clinton's 2016 presidential run

"*Terms of Disservice* bravely goes after the root of all internet evil—the consumer surveillance business model, a dire threat to our privacy, democracy, and way of life."

—Bruce Reed, Deputy Chief of Staff to U.S. President Joe Biden

"Lively, visual, and relentlessly clear-eyed, *Terms of Disservice* offers a disturbing evidence-backed portrait of a digital economy now cursed by catastrophic success."

—Jonathan Zittrain, George Bemis Professor of International Law, Harvard Law School

"If you want to understand how and why disinformation has swept the globe, imperiling democracies and encouraging mob violence, read this book. With engaging anecdotes and searing analysis, Dipayan Ghosh turns his insider's experience into exactly the kind of diagnosis and policy recommendations we desperately need. If there's just one book you read about the power of social media corporations over our lives and politics—and the new social contract needed to reform them—let it be Ghosh's phenomenal *Terms of Disservice*!"

—Dana Priest, Pulitzer Prize-winning *Washington Post* reporter and Professor of Journalism at the University of Maryland

"With so much disinformation online today, users might forget the Internet was meant to be a democratizing force. *Terms of Disservice* answers how we got here, serving as a detailed blueprint of the strategic moves made by Facebook, Twitter and Google over the last ten years and their ongoing influence reshaping societies and political discourse across the world. A former Facebook policy adviser, Ghosh's insight, paired with extensive expert testimony, offers readers a clear understanding of the most complex issues facing social media companies today."

—Courtney Norris, National Affairs Reporter and Producer, *PBS NewsHour*

"*Terms of Disservice* provides a disillusioned analysis of our dystopian relationship with big tech. Some of the news stories and anecdotes will be familiar, but this sweeping book strings them together brilliantly to expose the market, regulatory, and algorithmic forces buried underneath. And just when you think there's no way out, author Dipayan Ghosh offers a hopeful path forward."

—Shachar Bar-On, Producer, *60 Minutes*, CBS

"Ghosh's book, enriched by his real-world experience in tech and government, could not be more timely as the out-of-control internet increasingly enables anti-democratic, anti-competitive and other destructive forces. It should be priority reading for anyone looking to understand the trends at work and how social media can and must be reimagined to ensure both a vibrant global economy and the common good."

—Carol Giacomo, journalist and former member of *The New York Times* Editorial Board

TERMS OF DISSERVICE

How Silicon Valley Is Destructive by Design

DIPAYAN GHOSH

HARPER BUSINESS

An Imprint of HarperCollins *Publishers*

First published in the US in 2020 by
The Brookings Institution
1775 Massachusetts Avenue, N.W.
Washington, D.C. 20036
www.brookings.edu

First published in India in 2021 by Harper Business
An imprint of HarperCollins *Publishers*
A-75, Sector 57, Noida, Uttar Pradesh 201301, India
www.harpercollins.co.in

2 4 6 8 10 9 7 5 3 1

P ISBN: 978-93-5422-543-7
E ISBN: 978-93-5422-544-4

Typeset in Whitman

Printed and bound at
Thomson Press (India) Ltd.

MIX
Paper
FSC FSC™ C010615

This book is produced from independently certified FSC™ paper to ensure responsible forest management.

For my grandparents

Contents

Acknowledgments xiii

Preface to the Indian Edition xix

Foreword xxiii
JOHN PODESTA

Introduction 1

ONE
The Business Model 23
Data, Algorithms, and Platform Growth

TWO
Data 65
The Harvesting of All Knowledge

THREE
Algorithms 127
The Commercialization of Bias

FOUR
Platform Growth 163
Capitalism Consuming the World

FIVE
A New Social Contract 199

The Case for Radical Reform 249

Notes 253

Index 277

About the Author 297

Acknowledgments

This book is the culmination of a progression in thought I have made over the past decade. It's been a journey through which I've gained the privileges of becoming both independent and knowledgeable about the tensions between commerce and consumers that have become so central to the digital economy. That progression is one that has been propelled by countless friends, who along the way have taught me new things and helped me see things in new ways. I am indebted to all of you.

Terms of Disservice came to be because of my tireless and benevolent editor, Bill Finan. He gave me an incredible opportunity. Handing an author a limitless pen with complete intellectual independence is not easy. I owe Bill profuse gratitude for seeing a glimmer of something in my arguments and affording me the chance to share my perspective. The Brookings Institution Press team has been truly generous in making this happen. I owe endless thanks to Cecilia González for her sharp edits, Elliott Beard for his imaginative artwork, and Robert Wicks and Steven Roman for their incredible support.

My perspectives on technology have been heavily influenced by countless friends and colleagues. I owe deep gratitude to my colleagues from the Obama years. R. David Edelman, my most influential mentor from that period, helped shape my understanding of government service and the subtle force of bureaucracy. David's brilliance as a policymaker, thinker, and writer transformed my views of the world and inspired me to serve the Obama administration in the best way I could. Kumar Garg's unbelievable kindness in teaching me so much about the art of doing policy—quite often very late into the night at his office at the White House—was a genuine gift, with no return expectations and for which I will always be deeply grateful. His brilliance in generating truly novel ideas and making them happen time and again was a joy to watch.

Tom Power has a knack for bringing people together toward a consensus policy idea—and he possesses the brilliance to back up and enforce that consensus. I have learned so much from him. DJ Patil, Nicole Wong, and Tom Kalil gave me insight into how new technologies were shaping the world, and what the policy world could do to unlock even greater potential. Todd Park, Megan Smith, Alex Macgillivray, and John Holdren gave me the opportunity to pursue my work in government. I will always be grateful to them.

I also wish to thank John Podesta for his incredible leadership. Without him, none of our work to advance internet consumer interests in government—which shaped my perspectives on the nature of the national economy—could have happened. What I am continually struck by is his keen sense of morality in the scope of policymaking. Unerringly he thinks first of the little guy's circumstances in determining the impact of any policy. Simply put, that's just the way it should be.

A deep thanks to my many friends at Facebook, particularly the brilliant Steve Satterfield, Brian Rice, and Rob Sherman, who gave me a chance to see the world's most incredible internet company from up close. I learned so much during my time there. Kevin Martin, Will Castleberry, Erin Egan, and Joel Kaplan were all generous mentors whose leadership I deeply admire. Dan Sachs is a great friend and the most effective political and policy expert I know in the industry. The way he brings together his tremendous legal expertise with his eye toward getting things done is remarkable to watch. My deepest thanks also to Norberto Andrade, Ben Bartlett, James Hairston, and Andy O'Connell for your collaboration and friendship.

My colleagues at New America have left a deep impact on me. Bobby McKenzie and Josh Geltzer, thank you for your friendship first and your partnership in so many things. It has been a pleasure to tackle some very difficult policy issues with you. Thank you also to Viv Graubard, whose work has been an inspiration and whose support has always been incredibly thoughtful and thorough. I greatly appreciate the opportunities Kevin Bankston and Sarah Morris gave me at the Open Technology Institute, and am truly thankful for the guidance and support of Anne-Marie Slaughter, Cecilia Muñoz, and Peter Bergen.

Ben Scott, a special thanks to you. I deeply admire your ability to close political divides, develop truly brilliant policy ideas, and bring them to life. Our partnership in many endeavors—and your mentorship throughout—has helped me to achieve a clarity of mind and realize the path I should take.

Thank you also to James Timbie of Stanford University's Hoover Institution and Sam Sadden of Competition Policy International, whose invitation to write papers and editing guidance is much appreciated. Some of that work is published in this book.

My foray into technology policy began with my doctoral adviser professor Stephen Wicker, the brilliant computer scientist and information theorist. His powerful impact in many academic fields is undeniable. But it was his expertise in privacy that drew me to join his research lab and study the technical mechanisms by which the corporate sector should address privacy. The academic mentorship of professors Dawn Schrader, Lawrence Blume, Timothy Mount, Lang Tong, and William Schulze at Cornell University and professors Deirdre Mulligan and Shankar Sastry at Berkeley has too been deeply impactful on me.

Professor Peter Luh at the University of Connecticut gave me research opportunities that I will never forget, and I owe him my deepest gratitude. I also wish to thank professors Krishna Pattipati, Ali Gokirmak, Helena Silva, Rajiv Bansal, Faquir Jain, and the late Marty Fox. Faculty at the Massachusetts Institute of Technology, particularly professors Dean Eckles and Gary Gensler, have helped me understand a great deal more about the world.

There are many more friends from professional and social circles whom I wish to thank: Randolph Adams, Babatunde Ayeni, Paul Barrett, Amit Chandra, Amitava Chatterjee, Colleen Chien, J. Michael Daniel, Monica Desai, David Eaves, Layth Elhassani, Nathaniel Gleicher, Lindsay Gorman, Chris Herndon, Payton Iheme, Vijeth Iyengar, Will Johnson, Sina Lashgari, Jacob Leibenluft, David Litt, Nate Loewentheil, Laura Manley, Hannah Merves, Robby Mook, Greg Nelson, Catlin O'Neill, Patrick Parodi, Hetul Patel, Matt Perault, Mathias Risse, Larry Rohrbaugh, Alec Ross, Gail Roy, Ari Schwartz, Anthony Tre Silva, Nick Sinai, Jonathan Spalter, Prerna Tomar, Chris Weasler, Nancy Weiss, Darrell West, Secretary Ash Carter, and Secretary Cameron Kerry. Thank you all for your support of my work. I wish to especially thank Tara Kheradpir for her thoughts on the cover design and Jacob Beizer for his smart portraiture. Countless friends, particularly from the Sloan School of Management and Kennedy School communities, have supported me throughout as I have worked on this book. There are too many to name here, but you know who you are.

My incredible partners in crime at the Digital Platforms & Democracy Project—Tom Wheeler, Phil Verveer, and Gene Kimmelman—have been absolutely integral in my understanding of the modern technology economy. They have taught me so much through our innumerable conversations about policy, politics, and the regulation of new media. I am very lucky to be in their company. They have encouraged me at every stage of this work and have influenced my thinking about what is possible through the practice of developing internet policy. Here's to the four horsemen!

Meanwhile, our project at Harvard would not exist without the deep insight and thoughtful guidance of our bosses, Nancy Gibbs and Setti Warren, who have shared so generously with us. Their leadership of the Shorenstein Center on Media, Politics and Public Policy at the Harvard Kennedy School has made our work possible. My deep gratitude also goes to Nicco Mele, former director of the Center, who at every level has gone out of his way to support me. I am indebted to him as I am to Nancy and Setti, particularly for creating the space for my colleagues and I to be truly independent. I am equally indebted to professors Jonathan Zittrain and Urs Gasser of Harvard Law School for their generous support of my work and many other initiatives I have sought to undertake with their thoughtful guidance.

My mother and father have given me everything I have, and have both dealt with two unruly kids and treated us with equal love and respect. I don't have words to convey my appreciation. My grandfather, Nityagopal Ghosh, has been an indelible positive force in my life. His nature was irreproachable; as far as I could observe, he was well-intentioned in everything he did and every philosophy he espoused. Though he passed at ninety-eight last year and did not quite become the centenarian he always wished he would become, I hope that he will see that his good nature and positive spirit are contained in every breath of this book. A most special thanks goes to my brother, Debraj. He will be unaware of it, but I constantly look to his example for inspiration in my life. His genuine support, sharp wit, critical eye, and fighting spirit have been extraordinarily helpful to me through the years.

And of course, I had to save the most important for last: Loullyana Saney, you have given me everything in countless ways. I could never have gotten through these past few years—and written this book—without your endless support. I love you—and I am forever grateful to you!

Preface to the Indian Edition

When India banned TikTok last year, the government's calculus was apparent: protect the national security interests of Indian citizens given the combination of alleged personal data privacy risks concerning the TikTok mobile app and the possibility that its development firm, ByteDance, was to some unknown extent under the influence of the Chinese government. The underlying implication in the government's decision to ban the foreign application was potentially a stretch, but clear as day; there was intense concern among Indian policymakers and in the national security apparatus that China could gain access to the data held by TikTok on its users through its security authorities established in national law, apply machine learning to that information to infer the behavioral profiles of each and every user, including Indians, and use any resulting learnings about the Indian constituency to benefit Chinese short-run national interests.

The Indian government's fear was very real, particularly given the military tensions—and the loss of Indian lives at the country's borders in Ladakh—that the world witnessed in shock. When analyzed against this backdrop, the concern presented over TikTok is equatable to the very protection of soldiers' lives and the economic and security interests of the nation. This may be difficult to imagine at first; how can a mobile app and consumer privacy be linked to India's future prosperity? But when one draws the lines from the data exhaust generated by the company, its possible links to the government, and the sense that information is gold in today's geopolitical climate, these implications become resoundingly clear.

But our concern cannot be limited to just TikTok and its alleged privacy issues. India's ban of the app was executed on the basis of fears over app privacy and accompanying adversarial developments with China—but the harsh truth is that though other mobile apps and web services that we know and love have been brought up in regimes of regulated capitalism rather than centralized communism, the regulation of that capitalistic activity has, in the case of the consumer internet industry, not yet occurred. Facebook, Instagram, WhatsApp, Amazon, Google Search, Gmail—these are powerful corporations that are approximately and effectively beholden to no one in much the same way that the Chinese government too is beholden to no one. The commercial concerns presented by these services and many more at the very forefront of the consumer internet industry are substantial—and are already impairing democratic processes the world over. The collection of personal data on an uninhibited basis toward behavioral profiling; the development and application of sophisticated but opaque algorithms that curate our social content and target ads at us; and the intense engagement in aggressive platform growth tactics at the expense of the competition: these are the hallmarks of the modern internet industry. We can draw a direct line from these commercial practices to such democratic harms as the spread of disinformation, political conspiracies, and hateful conduct online—and yet we do nothing about it. We should. In fact, we must.

That is, at its core, what this book is about: the harsh economic realities we face today in the context of the sector of the economy that I know best, the internet. We live in an increasingly connected world—and this growing connectivity has an exponentially increasing impact on our experience of the world itself. Each year we are confronted with newer forms of technology: internet search, MacBooks, mp3 players, social media, smartphones, smart watches, e-readers, new-generation gaming platforms—the list will of course go on. This booming industry, which has blossomed in a vast green economic space untouched by regulation, has effectively been monopolized by a few powerful Silicon Valley firms. And though we cannot see it now, these developments will affect us individually, but perhaps even more crucial, they will impose critical taxes on the whole of society. Facebook, Google, Amazon and the like have chanced upon mechanisms to monetize our individual digital presences and reap all the economic benefits thereof for the shareholder class.

We must do something about this, we must protect ourselves and those around us from the overreaches of the internet industry. The question now is whether, as consumers, we will persist in this state of being harvested by the digital behemoths, or if we will stand up and take control of our economic rights. There is nothing wrong with capitalism so long as it is effectively constrained through meaningful, analytically rigorous regulation, and while the American way is to let the markets roam free, even they must be set to the side when consumer protection, human rights, and democracy itself are being systematically undermined. *Terms of*

Disservice analyzes how digital markets are currently roaming free, how the leading technology firms are exploiting that fact, and what we—the consumers—must now do about it.

When I first embarked on this project, I never imagined the scope it could and eventually did take, and for that I must thank the teams at HarperCollins Publishers and the Brookings Institution for giving me the ideal author's pen, one that carried no restrictions. This gave me the absolute license to lean into the analysis herein and deliver a cohesive narrative of how the profit machine behind consumer internet truly works. This book is informed by my analysis and learnings from my time working at Facebook and, prior to that, in the White House during the Obama administration. But it seeks to take us on a shared journey that all consumers encounter simply in the course of participating in daily life today.

New York City — Dipayan Ghosh
March 2021

Foreword

John Podesta

In the span of a single generation, the internet upended the American media landscape. The internet and social media held out the promise of accessing vast stores of information, building community, and empowering speech. Communications of all kinds were made easier by reducing friction, increasing access, and inducing transparency. The old broadcast model of selling the eyeballs of passive, mostly powerless consumers to advertisers shifted to an audience of amplifiers, editors, influencers, and content curators. But this rapid change with the internet and social media created new substantial threats to our privacy, our national security, and our democracy.

Dipayan Ghosh played a critical role in the American government's attempt to come to grips with those threats when he served in the Obama White House in the wake of the national security leaks by Edward Snowden in 2013. He was an invaluable contributor to a host of policy initiatives, including reforming the Electronic Communications

Privacy Act of 1986, advancing new encryption policy, regulating internet market competition, passing net neutrality rules, and implementing crucial economic diplomacy with the European Union.

In the post-Obama era, a new concern has come to dominate the field: the creation and dissemination of fake news. The sheer amount of false information that is spread by lightning-fast social media platforms means that internet media is nearly impossible to fact check. Information promoted by sources that look credible (and many that don't look remotely credible) spreads like wildfire. *Terms of Disservice* chronicals a new concern that has come to dominate the field.

The 2016 presidential campaign showed the spectacular dangers of the new media landscape to our democracy. Thanks to the dogged work of Robert Mueller, scores of investigative journalists, and the U.S. Senate Select Committee on Intelligence, we now have a good picture of what went on.

Russia deployed an eighty-person, multimillion-dollar digital campaign to undermine our fair and free elections by sowing division and spreading disinformation, in particular about Democratic presidential candidate Hillary Clinton. The Internet Research Agency, a troll farm with links to the Kremlin, used Facebook, Twitter, YouTube, and Instagram to incite and amplify chaos in the most aggressive election intervention by a hostile power in U.S. history. They sought to suppress left-leaning audiences, in particular targeting African Americans on Twitter and Facebook through pages and posts with titles like "Don't Vote for Hillary Clinton," "Don't Vote At All," "Why Would We Be Voting," "Our Votes Don't Matter," and "A Vote for Jill Stein Is Not a Wasted Vote."

Russian intelligence operatives hacked Democratic Party computer networks and the email accounts of Clinton campaign volunteers and employees, including, most famously, my own. Senior Russian intelligence operatives now under federal indictment stole hundreds of thousands of documents and published them under the cover of DCLeaks, Guccifer 2.0, and WikiLeaks. Special Counsel Robert Mueller, through federal indictments and his partially redacted report released in March 2019, confirmed that associates of Donald Trump's presidential campaign communicated with WikiLeaks and that WikiLeaks was a conduit for the Russian government.

Trump didn't just watch this happen. He actively supported the release of the hacked documents and emails, even inviting the Russians to steal Clinton's emails. He also went out of his way to call attention to WikiLeaks, mentioning the website's name 164 times—an average of more than five times per day—during the final month of the campaign, including during all three presidential debates. By late summer 2016, Trump seemed to have advance warning of what WikiLeaks would be releasing, as evidenced by his conversation with then-deputy campaign chairman Rick Gates, as recounted in the Mueller report.

This was an assault on our democracy that was perpetrated by the Kremlin and its cutouts and supported by the current president of the United States.

Political and social scientists are still arguing about the extent to which the Russian disinformation campaign affected the 2016 election. It seems obvious that voting (or not voting) under false pretenses—whether or not voters know they were lied to—is bad for a democracy. Additionally, it is clear that disinformation spread through fake news erodes the foundation of truth that social compacts are built upon.

What is inarguable is that once in office, President Trump began deploying a strategy that has been used by autocrats around the world—labeling anything in the mainstream media that is critical of him or his administration as fake news. The strategy is designed to completely confuse public perception. He's not just trying to spin the bad news of the day; all politicians do that. He seeks nothing less than to undermine the public's belief that any news can be trusted, that any news is true, and that there is any fixed reality.

The reaction of social media companies to the spread of disinformation on their platforms during and after the 2016 campaign wasn't just underwhelming but was also actively harmful to our democracy. Facebook, in particular, reacted to each new revelation with denial, delay, and dissembling to the American public. The company was slow to admit that the platform was used in 2016 by Russia and others to sow disinformation and even slower to admit its impact on the democratic process. Since the election, Facebook has made repeated promises to clean up the site but has taken only the most limited and ineffective steps to regulate the spread of disinformation through its platforms. After hiring a handful of editors to try to keep up with the deluge of disinformation, Facebook fired them and implemented an algorithm that, instead of limiting disinformation, immediately pushed it on to trending lists.

While testifying before Congress in April 2018, Facebook CEO Mark Zuckerberg promised to share the full range of the company's data with researchers so that they can study and flag the spread of possible disinformation on social media platforms. Eight charitable funders pooled $600,000 to create the Social Media and Democracy Research Grants Program to support researchers in analyzing Facebook's data. A year and a half later, most of the promised data has not been made available. The charities, frustrated with the lack of progress, have started to wind down the program.

So the problem continues into the 2020 election. Facebook, perhaps fearing the wrath of Donald Trump, has adopted a policy that it will not fact-check politicians' ads at all. In so doing, Facebook has created a safe harbor for any politician running for public office to lie with impunity—an approach no traditional print journalist or broadcaster would embrace.

The United States and the global community must take immediate action to protect citizens from the perhaps unanticipated, but now well-observed, negative effects of a Wild West approach to internet publishing and develop defenses against future challenges. While we know that disinformation affects the functioning of our democracy and can threaten human rights, there are sure to be more issues that emerge as the technology industry continues to evolve. As *Terms of Disservice* compellingly argues, we need to think about the consumer first through a digital social contract—a policy that attempts to level the balance of power between Silicon Valley, the U.S. government, and individual citizens.

TERMS OF DISSERVICE

Introduction

When Mark Zuckerberg took the stand before Congress to answer to the American people nearly a year and a half after Donald Trump was voted into the presidency, the public's great hope was that he would reveal specifically what had enabled the foreign infiltration of the national political process during the course of the 2016 elections.

What we received was quite different: an infuriating amalgamation of already public facts and obvious suppositions alongside essentially meaningless corporate commitments—all of which produced more questions than answers about the internet tools and technologies available to those who wish to raid our freedoms by influencing the natural march of political discourse in America.

Many long months after that initial Zuckerberg hearing, we are scarcely any closer to understanding the truth about Facebook or the broader industry it lies at the center of, the industry I call the "consumer internet." The cacophony of unfocused congressional inquiries, frustratingly misleading corporate gibberish, and public outcry that followed the Zuckerberg hearing and the many others before and since that time will have put any Silicon Valley executive at absolute ease: Washington, D.C., is apparently not ready to earnestly understand how the internet works and to regulate the companies operating over it to protect the American public from further harm. Instead, the companies leading the sector—Facebook, Google, Twitter, and Amazon—could very likely continue to operate with their existing business practices intact. It apparently matters little whether those practices imperil the public's interest.

But imperil the public interest they quite clearly do: inherently and almost directly so. Google, Facebook, and Twitter have each admitted that their platforms were systematically infiltrated by Russian disinformation operators in the lead-up to the election. The Russians' sole purpose was to subvert the American political process and inject a dose of chaos into our democratic discourse, thus unsettling the nation internally as well as our standing in the world and in the process feeding the personal and political hunger of the Russian prime minister.

By all accounts, Vladimir Putin was wildly successful in accomplishing this task. The result is an American president who failed to win the popular vote and quite possibly would have failed to win sufficient Electoral College votes, too, had he not received support from disinformation operators sitting behind computer terminals in the former Soviet Union—anonymous agents of the Kremlin who engaged in a coordinated communications campaign with a unified goal: to boost Trump's presidential chances by igniting a fire under the reputation of Democratic opponent Hillary Clinton.

I did not have to take my eyes from the screens draping the walls of the Javits Center in Manhattan that election night of November 8 to feel the reactions of shock among my friends and colleagues standing around me. The disbelief in the air was palpable as Secretary Clinton succumbed in one swing state after another.

I stayed to the bitter end that night, in shock. John Podesta, the chair of Clinton's presidential campaign, took the podium to share a few words shortly after two thirty in the morning; the handful of loyal supporters who had stuck it out through the night represented a sorry sight. I left shortly thereafter: I had meetings in the Washington office of Facebook, my employer at the time, starting at nine in the morning, and I would have to begin the solemn drive south right away to arrive on time.

The drive was long, and that night I was discouraged and disillusioned. But it gave me ample time to ponder what the world had just witnessed. I could not read any expert opinions or conduct my own analysis on the election results before the drive, but as I concentrated on the road the conclusion I came to was this: an insidious but coldly logical commercial regime underlies the internet that propelled Trump to victory—and without the silent internet economy that had enabled the coordinated spread of outright lies against his opponent, the nation would not have been wrenched from its tracks with such force.

Nonetheless, the Russian disinformation problem was only the canary in the coalmine signaling deep-rooted problems at the heart of the internet. The commercial regime behind consumer internet platforms—and the journey to comprehend it in its entirety—is the focus of this book. I aim to cut through the internet's weeds and depict its inner economic logic, as well as the economic factors motivating the decisions made behind the veil of Silicon Valley. And I do so in hopes that we can finally correct the course of a modern media ecosystem that has wholly failed the American people—and the global citizenry—time and again.

We have seen enough. It is time to repair the mass economic exploitation of American consumers that has taken hold at the behest of the technology industry.

The Internet as the Engine of Globalization

If you were to visit Calcutta today, you would be struck by the city's great divide. On one side are the rich colors of an established cultural identity. Wide, hopeful avenues like Chowringhee, Prince Anwar Shah, Rashbehari, and Central evoke a certain nostalgia for decades past. Sun-filled streets go sleepy by day but come alive by night, with the constant activity of tea stallers, coconut merchants, and shoe shiners on every corner. Romantic alleyways unite and diverge again in hapless disorder, lined with wall-to-wall flats covered with weathered paint of every stripe of the rainbow. The Victoria Memorial, the High Court, the Great Eastern Hotel, and the Eden Gardens cricket ground, remnants of a colonial India during which the city enjoyed an economic importance and accompanying panache that it has failed to regain over many long years, grace the city's center with elegance but serve as a constant reminder of decades of subjugation at the hands of the British Empire. And a people that has yielded the likes of the first Indian Nobel Prize winner and the first Indian Oscar-winning director today looks mostly to India's past accolades, excessively proud of the country's intellectual heritage and paying little heed to its place in the world's future.

In the other half of the city—literally so, from a geographic perspective—looms an imposing new sector that has cast away Calcutta's past and embraced the connected global economy in all its glorious commercialism. The new dedicated technology district sprawls over the northeastern edges of the city, flowing outward and upward as the tide of commerce grows. It only expands, graced by companies such as IBM, Infosys, Wipro, and Tata, all of which are engaged as technology consultants to the most important American and other foreign corporations.[1] The sector—known as Rajarhat—boasts towering campuses that take advantage of the enormous local pool of high-skilled young talent at a cost of technical labor that clients in the rest of the world find relatively low. It is especially well connected, with newly renovated highways connecting it to the downtown area and the nearby airport, and set for a major modernization itself. It is this industry more than any other that hundreds of millions of young people in India strive to be a part of because of its pay and prestige.

Calcutta—like many overflowing metropolises in the developing world—finds itself stretching its cultural fabric between two realities. The former is the soporific postcolonial world that places outsize importance on the city's high cultural heritage and intellectual contribution to the classical world. It is founded on the inherent brainpower of the local intellectual collective, a stronghold for the creation and study of literature, religion, science, and economics. Images of its greatest son, the writer and musician Rabindranath Tagore, grace the walls of the wealthy and the downtrodden. Pictured next to him is Subhas Chandra Bose, the nation's greatest martial leader during the era of independence, whose contributions have been palpable to every generation since the colonists departed. These two individuals represent a revolution of ideas and action that fueled the freedom movement for the subcontinental race a century ago.

But Calcutta's new world pulls the heartstrings of the local population in the opposite direction. Since national independence, the city has witnessed decade after decade of regional economic failure owing in large part to the failures of local governance advanced by a succession of communist leaders in the state, which have reduced Calcutta from what was perhaps the country's most prominent city to a relative trash heap when compared with India's booming metropolises situated to the far west and south of Bengal. This was no coincidence; there was a fair amount of political trickery behind the city's turn to communism. An influx of immigrants from present Bangladesh, an ethnically Bengali part of India before the partition, left the city overpopulated and its people disoriented. Leader after leader, armed with communist rhetoric, fought off would-be industrialists, citing the interests of the displaced poor for their decision to hold Calcutta back from engaging the global economy while the rest of the country accelerated forward—fast.

And so, in a world that became increasingly cutthroat and capitalist, and increasingly desperate, there was no place for the intellectual Bengali. Year after year, regional ministers fought off industrial activities, settling on hackneyed counterarguments: if the multinationals build near the tracks, where would the thousands living in the slums go? These arguments, fair enough, were not met with new policies that could uplift the poor. As time wore on, the question of industrial investment and entry became so rigidly politicized along party lines that nothing happened, and seemingly nothing could happen.

But then came the World Wide Web.

It was this—the advent of the internet—that finally managed to defeat the city's resistance to commerce. The industry did not have to work through municipal bureaucracy. There was so much money to be made that it could work around the local government completely if need be. As the modern world digitized commerce with the spread of the internet and increased use of the web, the market began to recognize that India could serve as the ideal technical support hub: millions of young, talented computer scientists, low marginal labor costs, and an English education system meant that this was a place teeming with economic opportunity for the Western barons of global digitization. India—powered by theretofore economically stagnant population hubs like Calcutta—could grease the commercialization of the internet along with all the background support the barons could want. And indeed, the barons have made their hay while the sun has shined. Thus was born Rajarhat, representing Calcutta's principal wave of industrialization since the debilitating world war and national breakup.

The city has profited immensely in the process, at least in the short term. If the technology hub did not exist, what would its young people do besides serve the local consumer market in more mundane ways or leave the city? The internet has revolutionized the way Calcutta thinks about industry. What was once an overpopulated consumer market has turned into a haven for business—even if less so than other major Indian cities, including Hyderabad, Bangalore, Mumbai, and Chennai—and this will drive the city's future as more young talent is attracted to careers in the technology sector and beyond. The industrial investments made in developing Rajarhat have afforded an economic energy that the city has not been party to in ages. To the same extent that there has always been an implicit pride in the historic nature of the city's many architectural establishments and intellectual accomplishments, there is a newfound bated anticipation over the economic possibilities Rajarhat promises to afford in perpetuity.

This new wave of activity has further impacted the region's cultural identity and political economy. Where once politicians used to beat back the efforts of the industrialists under the false platform of advancing the interests of the poor, they now claim credit for the economic benefits Rajarhat has brought. Commoners of the city now proudly boast the expansion of the northeastern sprawl, too; this new economic activity has positively influenced the region's outlook, bringing a fresh wind of activity.

The circumstance of Calcutta teaches us an important lesson: the internet—and, more basically, connectivity—will persistently find ways to break inefficient and ineffective institutions; it is in the internet's nature to tear down artificially imposed barriers by reaping the benefits of its low transaction costs and minimized frictions, challenging aging methods, industries, governmental regimes, and political systems in the process. This is the power of networks that Joshua Cooper Ramo has elucidated.[2] In Calcutta, the commercialization of Rajarhat was the result of an organic economic interest—a steady but powerful wave of investment that came on the heels of an industry recognizing that a real economic opportunity was close at hand. And the fact that these changes have brought with them long-desired reforms to governance is a welcome effect.

The story of Calcutta can be generalized to second- and third-world cities across the globe; the capacity for networked communication has enabled economic activity that amounts to more than just fortunes: for cities like Calcutta, it represents hope for future economic success in an ever-globalizing world.

The Dark Underbelly of the Internet: A New Vector for Public Harm

The internet has had outsize impact on the global economy and, because of its democratizing force on the spread of information, humanity's aggregate cultural richness. Indeed, its contribution to economic growth seemingly will only continue to grow. On the subject of international trade, John Stuart Mill wrote that "it is hardly possible to overrate the value, for the improvement of human beings, of things which bring them into contact with persons dissimilar to themselves, and with modes of thought and action unlike those with which they are familiar. . . . There is no nation which does not need to borrow from others."[3] As much could be said about sharing culture and information for the benefit of humanity; the internet has enabled us to communicate seamlessly for the first time in history. I have within my reach a wide selection of services I can use to transmit any data format at a speed and level of efficiency unthinkable just a few years ago. To say that this newfound capacity unlocks untold riches is an understatement.

But we must now assess the price of those riches. Recent years have seen a vast number of shocking incidents in the physical world enabled and facilitated by the rising use of digital platform services operated over the internet—incidents that extend far beyond the concerns around foreign influence in the course of American elections. In this section we discuss the emergence of these new harms.

Hateful and Violent Conduct

The atrocities against the Rohingya people of western Myanmar are front and center of these concerns. It was estimated in 2018 that Myanmar's military and local Buddhists killed no fewer than 24,000 Rohingya people—a chilling number that the United Nations has unequivocally said constitutes genocide. In their review, UN investigators noted that many of the Myanmar officials implicated in the genocide charges had been using Facebook to disseminate hateful content to fuel the killings, rapes, and beatings of the Rohingya. What followed was a hard slap to the company: the UN noted that "although improved in recent months, Facebook's response has been slow and ineffective. The extent to which Facebook posts and messages have led to real-world discrimination and violence must be independently and thoroughly examined." Facebook responded quickly, noting the obvious: "The ethnic violence in Myanmar has been truly horrific. Earlier this month, we shared an update on the steps we're taking to prevent the spread of hate and misinformation on Facebook. While we were too slow to act, we're now making progress—with better technology to identify hate speech, improved reporting tools, and more people to review content."[4] With the delayed takedown of eighteen Facebook accounts and fifty-two Facebook pages, along with the presumable addition of some new staff who would monitor the company's business in Myanmar and attempt to obviate another human rights disaster, the company seemingly felt it had washed the grime from its hands.

Populism and Radicalization

In another category, internet platforms have abetted the new prominence of populism, which has dramatically been on the rise not just in the United States and Europe but across the globe. The Rohingya genocide is one instance, but at another scale is what has happened in Brazil, where Jair Bolsonaro was elected in 2018 under the banner of the right-wing Partido Social Liberal.[5] Bolsonaro maintains views ranging from the highly reprehensible to the deeply unsettling, including that women should not receive the same salaries as men because they become pregnant and therefore diminish work productivity;[6] that "the state is Christian, and any minority that is against this has to change";[7] and that "if your child starts to become like that, a little gay, you take a whip and you change their behavior."[8] While some of his most eye-popping positions have been explained away, what Bolsonaro stands for is clear. More recently, his radical views have gone a significant way in exacerbating Amazonian deforestation.[9]

But what is also becoming increasingly clear is that YouTube propelled him to the presidency. Brazil, a remarkably diverse nation with significant population segments that have ethnic roots in Africa, Europe, and precolonial South America, is addicted to YouTube. Technologically speaking, that is more a natural phenomenon than a negative aspect. Google's industry-leading video platform has become the go-to service for any topic under the sun—sports, history, music, movies, and all else—including, unfortunately, radical political content. Many young Brazilians have spoken openly about how YouTube was centrally responsible for their radicalization—expressed eventually through the ballot box with votes for Bolsonaro. The science appears to back up their claims about the platform. Researchers in Brazil and at Harvard University analyzed 331,849 videos and more than 79 million comments on the platform and studied a multitude of content pathways over the platform, entering common search phrases on YouTube, selecting top recommendations from the search results, and ultimately seeing where those recommendations took them.[10] The answer: straight down the rabbit hole of far-right conspiracy. After leading a user to any video in the realm of politics or entertainment, it was often the case that the user's pathway would run into far-right channels, and furthermore, once a user watched one far-right video, the platform would often recommend more. Bolsonaro was one of those conspiracists, and many Brazilian youth have described how they grew addicted to watching his videos and those of his followers.[11]

Other regions of the world have experienced similar outcomes, with platforms such as YouTube predicting citizens' political preferences—perhaps incorrectly—and aggressively pushing users into compelling but hateful trails of conspiracy and radicalization. Over the many days I have spent watching videos of alt-right activists—Jared Taylor, Mike Cernovich, Stefan Molyneux, and, most engrossing of all, Richard Spencer—I grew to understand the likely impact of watching hateful political communicators preach radical ideas.

Localized Misinformation to Incitement

India has witnessed the spread of violence, particularly against the country's most marginalized communities. There is no better example of an internet platform service derailing from its intended operation than the use of WhatsApp to spread hateful lies and conspiracies against targeted individuals.[12] At issue is the use of encrypted messaging to large groups that has been used by various propagators throughout India to associate certain targeted individuals with rumors of child abduction and organ harvesting. As the messages have circulated, various groups have been incited to commit violence against the identified targets, including a series of local lynchings.[13] The practice of sharing encrypted messages—such that hateful conduct becomes difficult to detect by the platform service provider or even governments—has been taken up in national politics in India as well, with the right-wing Bharatiya Janata Party reportedly initiating hundreds of thousands of WhatsApp groups targeted at ultra-local communities in India to automate the spread of propaganda throughout the nation.[14] While some might contend that the WhatsApp service does not implicate public interests and is not aligned with the business model pursued by Facebook (which owns WhatsApp), there is a direct line that can be drawn from WhatsApp's operations and Facebook's profitmaking pursuits, as I will discuss in chapter 2.

The Diminishment of the Fourth Estate

The prevailing digital platforms—especially those operated by Facebook and Google—have unconditionally diminished the importance of the organizations and individuals that have traditionally reported the news. Journalism today is floundering in the United States and in many other localities around the world where social media applications have become popular.

Breaking down the meaning of and implications surrounding journalism's demise is critical. Many news organizations have been put in financially precarious positions in recent years, especially local print newspapers large and small, including the *Tampa Bay Tribune*, *Pittsburgh Tribune-Review*, *Seattle Post-Intelligencer*, and others that once enjoyed circulations in the hundreds of thousands.[15] This has given rise to "news deserts," pockets of the country that lack access to a print publication and therefore a medium by which to get news.[16] Further, it has meant that the industry of journalism has contracted; beyond the job consequences of the aforementioned shutdowns, many organizations that remain in service to their constituencies have nevertheless had to cut staff. As Jon Allsop has discussed, 2018 was an especially bad year for the aggregate job count in journalism.[17] Meanwhile, the only traditional news organizations that remain healthy to at least some apparent extent—like the *New York Times* and *Washington Post*—walked into the digital age with tremendously powerful brand names that extend their reach well beyond their relative localities. The growing number of industry barons like Jeff Bezos, Laurene Powell Jobs, and Marc Benioff buying up famous news organizations that have traditionally functioned as self-sustaining businesses does not bode well either.

Of course, we could interpret the decline in the number of traditional news organizations as a simple effect of fluid American capitalism at work, an instance of Schumpeter's "perennial gale" of creative destruction featuring a cycle of innovation and disruption.[18] As society has evolved and more and more people have gravitated from traditional news media to social media and other digital platforms to read current news, traditional news media has inevitably struggled to attract adequate advertising from marketers to support the sustenance of the business, causing them to contract or even shut down. But such a mass contraction has consequences. As more newsrooms shutter and the people in those that remain wane in number, the practice of journalistic inquiry will increasingly suffer. Fewer trained journalists will be in the field acquiring and investigating the hard facts of the Russian disinformation operations or even the health of local businesses on Main Street. Meanwhile, we as consumers will source our information from digital platforms, which populate our feeds with news and social posts—a circumstance that comes with its own vast set of problems ranging from the economic to the political.

The connections to the actions of Silicon Valley internet firms are not only many but also difficult to discern. Clearly, though, there is a deep-rooted societal problem underlying the predictable transition of ad dollars from traditional news media to digital platforms. As traditional news organizations have declined in numbers and revenue, citizens have collectively relied on the internet for a greater share of news facts—but those facts have been riddled with lies, hatred, and conspiracy theories driven by those who have interests in disseminating such content. Meanwhile, the platform firms have generally been standoffish about the need to regulate content until their backs have been pushed to the wall. They have furthermore been forced into a dispute with the news industry and others over whether they should simply be considered indifferent internet platforms that are not liable for the veracity of the content they spread—a standard that is protected by the now-infamous Section 230 of the Communications Decency Act—or whether they should be considered media outlets like the traditional news industry is and as such be regulated by the federal government.

The fact that digital platforms divide society into population classes for ad targeting purposes and curate content on a personal basis for individual users should suggest the correct answer: their curation of social and news content, alongside their relative centrality in the modern information ecosystem, will necessitate media-like regulation. Such legislation is already on the table for consideration in Congress.[19] Passage, however, is controversial and likely will be drawn out over many long years. In the meantime, we must wait and see if the latest voluntary measures that companies like Google and Facebook have taken will amount to any meaningful change. Both have established well-funded news initiatives that nominally support the interests of journalists (although these projects have been vilified by some experts).[20]

Terrorism

One of the most terrifying incidents in recent times, once again involving Facebook, took place in Christchurch, New Zealand, in March 2019. Brenton Tarrant, a young Australian man from New South Wales who has been described as a far-right white supremacist, opened fire on the Al Noor Mosque and later the Linwood Islamic Centre.[21] Tarrant's chilling use of Facebook Live to stream his approach to the mosque and what he did inside it incited a global controversy over the use of online video-streaming services, including Facebook Live, LiveLeak, and YouTube, all of which were used to disseminate the shooter's video. YouTube product manager Neal Mohan noted that the "volumes at which that content was being copied and then re-uploaded to our platform was unprecedented in nature."[22] Prime Minister Jacinda Ardern, after copies of the video showed up in her feed well after the fact and despite Facebook's attempts at content takedowns, took the matter into her own hands, hosting talks with French President Emmanuel Macron to confront the ongoing social media crisis.[23] Many attended, including British Prime Minister Theresa May, Canadian Prime Minister Justin Trudeau, and European Commission President Jean-Claude Juncker, along with Twitter Chief Executive Jack Dorsey, Microsoft President Brad Smith, and Google Senior Vice President Kent Walker. One notable absentee: Mark Zuckerberg.[24]

Digital Commerce: The Thread Connecting It All

What is the thread that connects these terrible new circumstances? I would put that it is the economic logic—the business model—underlying the consumer internet itself.

Consider this possibility: the commercial nature of Silicon Valley, and more specifically the internet sector, is causing the sociopolitical problems that beset us today. The conjecture is not obvious; applying this lens to the Russian disinformation problem, to take one example of a public harm engendered by the internet, is not trivial. There are nefarious actors—people in the physical world—who push coordinated disinformation into American media markets to trigger harmful political effects. Presumably they, of course, are also to blame.

Industry executives such as Sheryl Sandberg, Facebook's chief operating officer, accordingly describe such disinformation propagators in clear and aggressive terms. She has noted that "there will always be bad actors, and I don't want to minimize that, but we are going to do everything we can to find bad actors," adding further, "we're not looking at these tradeoffs like 'oh, it's going to hurt our business. . . . People's trust is the most important thing."[25]

Sandberg places the blame for the spread of disinformation on one entity—the set of "bad actors"—and none on her firm. What the company knows too well is that a silent machine sits behind 1 Hacker Way's shining exterior and—like any other Silicon Valley behemoth—advances solely the long-run profitmaking interests of the company's owners and investors over any other consideration. And there is an implicit alignment in the commercial goals of Facebook and the persuasive goals of the Russian disinformation operators. Both desire the user's maximal engagement with the content at hand, and unless the appropriate rules and regulations are set for the platform, the company will promote whatever makes it the most money. The responsibility for the spread of disinformation must be shared by the entities that created it and by the entities that enabled its dissemination. We cannot ignore the economic alignment of their objectives.

If we, as common consumers, were to take our understanding of the disinformation problem exclusively from their statements, we would not suspect the industry itself as a primary culprit. But it would be foolish to ignore the possibility that this problem is a by-product of Silicon Valley's pell-mell pursuit of mammoth profit margins by expanding domination across the territory of the digital landscape with reckless abandon—or, to rephrase Facebook's old motto, by "moving fast and breaking things."[26] After all, we did not have this problem before the age of vast internet commercialization; we were not plagued by misinformation and foreign election interference in the ways that our political communication networks were infiltrated in 2016.

It is on us to dissect what internet executives have thus far said and take it for what it largely is: commercial propaganda. The reality is that the economic infrastructure that defines the internet is centrally responsible for the widespread damage that has been done to the American media ecosystem.

Of course, as Sandberg suggests, nefarious actors are working for the Kremlin and beyond and have it in for us and are eager to develop new methods for injecting opportunistic political messages into the American discourse. In fact, we may never be able to make them disappear completely; even if the U.S. government were to organize a clever set of sanctions that debilitated Russia's intent and capacity to attack our political and information systems, other actors would quite likely spring up and aim to subvert the sovereign strength of the United States. We have already seen some evidence of information operations emerging from China.[27] That this is happening should come as no surprise; such is the way of a multiethnic world replete with transnational economic competition and the resultant geopolitical uncertainty.

But what did not exist until recently was the current mode of internet commerce—led by the likes of Google, Twitter, and Facebook—that has promoted the reemergence of digital propaganda and other classes of harmful communications. It is the creation and facilitation of the novel influence market hosted by the consumer internet firms that is responsible more than any other single factor for the prevalence of this series of public harms.

We need to step away from the industry executives' injections of engineered noise and distill a comprehensive policy regime—a new internet order—that can once and for all motivate a truly earnest conversation about what the United States and jurisdictions around the world should do to contain the capitalistic overreaches of the consumer internet industry.

Unveiling the Web of Secrecy behind Digital Commerce

To design the appropriate regulatory intervention to contain Silicon Valley's public harms, we must start with a thorough analysis of the business model that is in play. At its heart, that business model is joltingly simple: it involves the creation of compelling internet-based apps and services that limit competition over the internet, as well as the uninhibited collection of fine-grained information on individual consumers to create a behavioral advertising profile on them, and the ongoing development and implementation of a set of algorithms, including artificial-intelligence systems, that automatically curate social content to engage consumers and target ads at them in a programmatic manner.

My contention is that given this business model, we need aggressive, reformative policy regimes that can better ensure individual privacy, increase consumer transparency, and promote market competition in the consumer internet sector. It is only when examined through the lens of the consumer internet's business model that the purpose of the industry's corporate decisionmaking comes to life, enabling the type of truth-busting skepticism necessary to keep this industry and its executives honest.

Why was the Cambridge Analytica incident—in which Facebook data pertaining to some 87 million users was compiled and illegally sold to the British political consulting firm that was contracted to Donald Trump's presidential campaign—a critical event from a technical perspective? Was it really just like the data breaches involving Capital One or Target or Sony or the hacking of Experian customers' e-mail addresses and financial information? Resoundingly, no.

Experts have discussed at length the troubling capabilities that Cambridge Analytica advertised. It could, for instance, engage in psychographic analysis and categorize people along five psychological qualities—openness, conscientiousness, extraversion, agreeableness, and neuroticism—determined by their behaviors and activities as revealed through their Facebook engagement data, including "Likes." This is not a unique capacity. IBM, for instance, has used its Watson artificial-intelligence platform to generate detailed classifications of psychographic traits and apply them to text.[28] Nielsen has advertised similar capabilities. And, in fact, it has widely been suggested that Cambridge Analytica's capabilities in driving psychographic inferences on users for client applications in the political context was limited, at best.[29]

But beyond any capacity Cambridge Analytica may have had in developing psychographic conclusions about a given user, I think the most critical aspect of the incident involved the potential breach of personal identifiers, possibly including what are known as Facebook User IDs—individually identifying pin numbers that would enable whoever has the Cambridge data to create vast (or narrow) target audiences for digital advertising campaigns on both Facebook and non-Facebook platforms and to coordinate highly effective disinformation campaigns. Most advertisers on Facebook have to go through the hoops of mapping known customers to their Facebook accounts, which is a process that does not always yield many matches to Facebook profiles. However, if the Cambridge Analytica breach included user IDs, the mapping would have been done for them. In other words, instead of reaching a matching yield of just 30 percent or 50 percent, the firm's clients could reach 100 percent of the 87 million accounts whose data the firm had obtained. Given that the vast majority of the 87 million accounts were those of American voters, the firm essentially had access to a targeting-and-tracking regime via Facebook's advertising platform that would have enabled a degree of illicit targeted political communications the likes of which we have never seen before.

While we might have some clues from independent researchers about what data may have been a part of the breach given analysis of data generated by Facebook's old application programming interface (API),[30] the company has provided very little in the way of public guidance to address the actual data exposed in the breach. I have yet to see an in-depth analysis of the Cambridge Analytica incident that highlights the dangers associated with the breach of personally identifying Facebook data that potentially included user IDs—despite the reality that this likely is the single-most critical facet of the Cambridge Analytica incident.

That Facebook consistently fails to address the implication that specific users could readily have been targeted on the company's on- and off-platform advertising services using the identifying data obtained by Cambridge Analytica is highly discouraging. It composes precisely the sort of obscurantism that deserves great public scrutiny and thorough investigation. Consider, also, Facebook's recent decision to enable texting across its three major internet-based text-messaging services—Instagram, Messenger, and WhatsApp. Each enjoys hundreds of millions of users, which for Messenger came about in large part because Facebook some years ago forced many existing users who had Facebook accounts to download Messenger if they wanted to text over Facebook with their Facebook friends. On its surface, Facebook's decision to enable this cross-messaging might appear to be a boon to consumers—who would reject the efficiency and ease of maintaining contacts on a single service versus switching back and forth between apps to keep in touch with friends?

However, what I find most striking about this decision is the way it is presented to the public by the company—as a technical change that will offer great convenience to customers. Of course, this might be one of the factors that drove the company to make this change; there was every possibility that by integrating the three applications users would find their experience to be more convenient and would therefore communicate with each other to an even greater extent over the three services or save some time in their day by not having to switch between mobile applications and websites.

But if everyone who uses internet messaging is a user of any of the aforementioned three apps and therefore can message with anyone else who is also a user of one of the three networks, what could drive the scaled adoption of a new texting service that might someday compete with Facebook's universe of texting networks? As such, this advent would also accomplish something starkly new: propel Facebook along its already-secure path to monopolization of the internet-based text-messaging sector. With this artificially imposed and potentially unfair market strategy, Facebook may at once shut out the possibility for any rival to compete with its apparent takeover of the messaging industry and stave off potential threats of antitrust regulation by claiming that de-integration of the apps would in the short term be hugely taxing to end consumers. It is predictable that the company would argue against the notion that it is monopolizing the industry. But I believe this strategic move to consolidate the market is significant and potentially warrants the application of robust competition rules to maintain the vibrancy of internet services and economic fluidity in the texting market.

These examples suggest a silent, commercially driven insidiousness at play behind the exteriors of the biggest Silicon Valley internet firms. To be sure, they are not alone among firms in the U.S. economy attempting systematic exploitation of American consumers.[31] All firms attempt it—that is the nature of our economic design. But there can be no doubt that significant regulation of the internet industry is needed at the earliest possible political opportunity, or else we will risk the failure of democracy in favor of the fortunes of a few: the internet barons. There come times throughout history when technology implicates public interests, and in those times we must for a time encourage the law of the suppression of radical potential.[32]

The internet has been an incredible force for good. The impact that Facebook and Twitter have had on the people of Tunisia and Egypt nearly a decade ago, as the world watched their protests, is a remarkable combination of positive political upheaval and radical economic enablement. Imagine those who experienced the Arab Spring; for the first time in their lives, millions felt intellectually liberated from their governmental oppressors. The aggregate mental release sent shockwaves around the world—and particularly in the Arab world, where many continue to experience turmoil and oppression every day but now have a clear example of what freedom can bring as well as what can be done to achieve it. Social media platforms were instrumental in bringing these positive changes about, as the government's power to suppress was shaken by the will of a populace. Wael Ghonim, a friend and perhaps the most important organizer of the movement in Egypt, remarked to CNN that "this revolution started online, . . . it started on Facebook," adding that "if you want to liberate a government, give them the internet." By enabling new forms of democratic process, the social media firms have also facilitated progress to counter some of the world's long-standing injustices.

That said, the focus from here on is the direct and immediate harm that the internet has wrought around the world. The coronavirus (COVID-19) pandemic is the most recent, high-resolution illustration of these harms. Most healthcare professionals and media outlets were quick to attempt to resolve points of confusion about the virus, but that did not prevent the tidal waves of misinformation and disinformation that spread over internet platforms and adversely affected an unknown number of people. Some of the hoaxes that have been spread through Facebook ads (including ones created by Consumer Reports to test the platform's ad review system through a bogus page for a made-up organization called the Self-Preservation Society) suggested that coronavirus is a scam and that social distancing was entirely unnecessary, when in fact this was the single tactic that made the greatest difference in preventing the spread of the virus.[33] Others suggested, by showing doctored lab test documents, that former Vice President Joe Biden had contracted the coronavirus.[34] Yet more hoaxes indicated that there would soon be food shortages in the United States.[35] The list goes on, to the extent that news outlets like NPR found it useful to host numerous segments specifically identifying and screening such misleading information.[36] (In the case of Consumer Reports's investigation, Facebook's system approved the ads, failing to identify any issues or potential harms. Of course, the organization pulled the ads before Facebook could publish them.)

The main culprit in the spread of misinformation about the virus has been social media, as has come to be the norm. The vast majority of the misinformation concerning the features and spread of the virus has occurred over platforms operated by dominant digital platforms. On a more novel note, however, much of this spread has occurred over messaging threads on WhatsApp, where fake rumors about steps that could be taken to protect oneself had been sent to large groups of people who were then encouraged to send them on to friends and loved ones. Facebook recognized this problem and in turn placed serious limits on the forwarding of such messages,[37] a tactic it has enforced in other past situations as well.

But will these kinds of corporate policies—narrow in their conception and application to timely problems—truly be enough to protect a society and the underlying media ecosystem we have worked so hard to guard and cultivate? Or could there be a deeper-lying economic demon within the internet industry that must be extracted and eradicated to diminish these problems in the long run?

Our purpose is to address these very questions. In this book, we will traverse the dark underbelly of the internet—the practices and positions that its leading lights are less proud of touting and that I believe have directly germinated the social harms we now witness, in the context of coronavirus and far beyond. Ultimately, we will also examine possible paths forward to mitigate these harms through progressive policy.

I believe that only with such a redefinition of our social compact with corporate America—and starting in Silicon Valley—can we again realize a world in which the internet is once more a universal gift to humanity.

ONE

The Business Model

Data, Algorithms, and Platform Growth

Civilizations become tied to ideals. As individuals, we become obstinately committed to common and widely accepted ways of thinking; we are inherently inclined to be mentally lazy, and nothing could be lazier than subscribing to rigid social convention. But as we take such convenient routes as individuals, collectively we risk intellectual stagnation. We risk failing to see the truth that lies squarely before us. With the publication in 1962 of *The Structure of Scientific Revolutions,* Thomas Kuhn shared a bold idea that stood the academic world on its head. Seemingly overnight, Kuhn revolutionized the way we would think about the very development of knowledge for decades to come—unsettling the time-worn paradigm of knowledge cultivation and transfer in the process.

The general perception at the time was that the fields of engineering and science, particularly in academia, were the ultimate keepers of knowledge. Society contended, perhaps reasonably so, that if there was anything about the physical world we might wish to learn about, it was to academics that we should turn. In the process, we would subscribe to their methods of teaching—attending universities and signing up for their classes, for example—to entertain hopes of learning neurobiology, or theoretical physics, or accounting. Think of an economics professor who presented his or her own ideas about how the monetary system should be governed and could accept no competing theory—especially a theory projected by a thinker who did not graduate out of the liberal economics tradition largely adopted in the field.

Kuhn contends that because of this mindset, academic contributions overall—such as those made by university professors—are often necessarily minor. Scholars build on a vast body of existing knowledge, such as neuroscience or finance, by tweaking the margins of an already-published theory or experiment and discovering some minor facet of reality that is, at best, a fringe contribution as far as markets are concerned. Kuhn's theory seems to resonate today: many brilliant students claim that they choose fields outside academia because the typical academic contribution carries little impact on the world outside the academic's career progression itself. Doctoral graduates frequently opt to work in fields in which they contribute to tangible outcomes rather than pursue the academic profession—or, for that matter, make a lot more money than an academic life can offer.

Kuhn further critiqued that low-impact ideas that by their nature will not significantly advance our understanding of the physical world are in fact the kinds of work that will most likely be published by peer-reviewed journals, because journal publications typically must be endorsed by other academics. Kuhn argues that this dynamic engenders a systemic dilemma whereby big ideas that could truly challenge traditional conceptual frameworks and revolutionarily expand our understanding of the world are unlikely to win the stamp of approval from the academic community. In fact, scholars who attempt to do so as academics might be harming their careers. To survive as an academic, Kuhn suggests, one must not stray from the community's unwritten rules—at least not too far.

At a minimum, Kuhn's argument makes a great deal of sense in American academic societies and in the scientific and technological fields. Young American scholars are educated through a grade system that is based on a clear hierarchy. They enter as inexperienced high school graduates and might choose to attend college to major in neurobiology or computer science—all to land a job at a technology company or bank. Some of us might decide that a bachelor's degree is not enough and choose to enroll in a graduate degree program. This might lead to doctoral research roles and teaching. Those who choose to extend their time in academia could later opt to engage in postdoctoral study, deepening their subscription to the academic hierarchy. And those successful in finding an opportunity to start a tenure-track academic role might ultimately choose that route, climbing the difficult rungs from assistant to associate to full professor and perhaps even department chair. To successfully navigate the field, particularly in engineering and the hard sciences, scholars must typically publish in the journals and conferences deemed by the relevant academic establishment to be adequately rigorous and competitive for the sharing of new research findings. Instilled in the academic practice, particularly in technical fields, lies an intense publish-or-perish culture.

But herein lies the real question that Kuhn asks us all to ponder: Do such academic contributions and the academic hierarchy they buttress really constitute the ideas that are of greatest novelty and greatest value to society? Possibly. But it is a process that remains opaque, with purview kept exclusively to the wisdom of the academic line of succession. To Kuhn, this setup represents a regime of power not particularly different from any other, but one that carries a fundamental flaw: complacency. What if the carefully combed and curated academic representatives of engineering and science, who over time had consistently subscribed to the idea that the traditional academic way is the only path to knowledge development, were fundamentally wrong about their theories of the world? What if they were missing entirely novel ways of thinking because they could not see past traditional, accepted frameworks? And, most critically, could there be other competing paradigms that deserve attention and analysis?

At the center of Kuhn's analysis is the story of the Copernican revolution, in which Nicolaus Copernicus posited a bold new idea: the Earth was not at the center of the universe, as the ecclesiastical order and Renaissance society had contended for many centuries. His ideas were accordingly shunned by the church and academic order, which persisted with the Ptolemaic model featuring a stationary Earth at the center of the skies. Copernicus's theory only started to gain sway when it became increasingly difficult to explain away discrepancies between Ptolemy's paradigm and new instances of real observations, particularly those of Galileo Galilei, Johannes Kepler, and Isaac Newton. Only after these scientists' ideas started to come together and force a divergence from the original path through hard science could the conversation around the physics of the solar system begin to move in the direction of truth. This was the Copernican paradigm shift.

In that instance, the development of astronomical knowledge was hindered by an academic hierarchy that had previously adopted Ptolemy's description as truth. Kuhn contends this was a theme of the academic field, to such a degree that it systematically obstructed understanding of the reality of nature. The result of his analysis is the idea that the academic world—and the people who subscribed to it—was unknowingly party to a cultural paradigm that might have been slowing humanity's quest for real knowledge; that because of the learning pathway moderated by the academic community, revolutionarily new knowledge that could change the world would seldom be shared. Academia had subscribed to a paradigm that favored conservative alignment over groundbreaking inquiry and truth, leaving the vast majority of people unable or unwilling to break from the present paradigm. We were minions living in an intellectual prison.

With this strange new theory, Kuhn sent shockwaves around the world.

The Internet's Intellectual Free Riders

As we approach the fourth decade of the consumer internet, I believe our society has adopted a new paradigm—one that could slowly constrain our intellectual future.

Since the dawn of the internet, we have revered it. Through the years we have thought of it as the technology that would revolutionize global commerce, democratize all communication, and topple political backwardness throughout the world. It was the enabler of vast economic opportunity in the developing world and the facilitator that would unlock the mass commercialization of industry around the world—starting with the United States. Indeed, our written history suggests it has been just that.

The problem is that while corporate evangelists highlighted the shining veneer of the internet to project its virtues and values to the industrial world, entrepreneurs discovered new opportunities to exploit consumers and tackle markets worldwide. They rode the coattails of the image of the internet—they were the new kids on the block, the college bros, the Wall Street bankers and venture capitalists who exclusively possessed the know-how to make something of this new media platform and its underlying power. We revered them all, including Bill Gates, Brian Chesky, Eric Schmidt, Evan Spiegel, Jeff Bezos, Jerry Yang, Larry Page, Marissa Mayer, Mark Zuckerberg, Peter Thiel, Sergey Brin, Sheryl Sandberg, Steve Jobs, Tim Cook, and Travis Kalanick, as well as those in the investment community, such as Mary Meeker and John Doerr. While they consolidated markets, hoovered personal data, and leapt over the consumer interest, they inaccurately projected that they and their companies *were* what we knew as the internet—that their businesses represented the novelty and wonder of the public domain that is the internet. It was the perfect way to brand the global conquest of the century.

The reality that has since emerged in place of that imagined magnanimity and ingenuity of the internet entrepreneurs is now clear:

they were just opportunists operating in an open greenfield of unregulated space, just like any Carnegie, Mellon, Morgan, Rockefeller, or Vanderbilt of the past. There was nothing special here. The new robber barons of the web projected the commercial propaganda that their earth-shattering products were good for everyone in society, just like the underlying free infrastructure that enabled their conquest in the first place: the internet. And that is the viewpoint that has largely prevailed over the past three decades: Facebook connects everyone, Google indexes all knowledge, and Amazon enables new markets. And all for free. That is the paradigm we are living in: consumer internet firms have done amazing things for the world, giving us the tremendous gift of connectivity. And even if this aura is diminishing in elitist circles, for the vast majority of internet users, it is not.

At first, the corporate evangelists offering these perspectives may have been sincere, even benevolent. They may even have thought they were true. The consistently earnest Tim Berners-Lee—who invented the World Wide Web, the technical protocol that we as consumers typically use to access our favorite websites—had grand ideas for what the internet could, and still can, accomplish. The internet in his view is the most sophisticated communications medium the world has ever seen, a system that enables the digitized communication between two terminals anywhere in the world. Berners-Lee envisioned the internet as an open space of limited governance, a decentralized forum for new ideas and protected communication, and he knew that if humankind could effectively organize it as such, it possessed truly remarkable capabilities for the benefit of us all.

This inspirational view was eventually subsumed by the businesspeople of the internet. What started out as a crude college dating service became for Mark Zuckerberg a new way to connect people around the world and allow friends to communicate with one another in the most seamless way imaginable—Facebook. For Eric Schmidt, the longtime head of Google, his company was the world's literal answer to collecting, analyzing, and presenting all of humanity's knowledge. Jeff Bezos's

Amazon had tremendous potential impact in its creator's view: it could and should become a meeting place for all of the world's merchants and customers looking for deals on anything from jumper cables to luxury cars. In each of these cases of corporations that over time have overtaken large segments of the global economy through cutthroat enterprise, the chief executive projected and perhaps even believed that he was creating something that was morally desirable and would benefit and uplift the consumer masses—all fundamentally incapable of stumbling on the exploitative business model at the core of the consumer internet.

Clearly, the case of Tim Berners-Lee is different from that of Mark Zuckerberg, Jeff Bezos, and Eric Schmidt. Berners-Lee is a scientist. The others are businesspeople. For Berners-Lee, the personal motivation to argue that the internet is a positive democratizing force comes from a place of intellectual integrity: it is the genuine sentiment of a rigorous thinker who earnestly considers the trade-offs that are inherent in such an intellectual position. For the others, there is a business interest arguing that the internet will be the world's panacea—and that it will be their companies over others that will resolve the world's problems. This disparity exists for the trajectory of the internet's broader conceptualization, too. We were once excited about its unique potential to connect two people on opposite sides of the world. It was an unbiased public space that carried an international brand of openness, optimism, and hope, where anything was possible and interaction truly had the ultimate opportunity for democratization, whether one was operating an oil rig off the coast of Bahrain or sitting in a basement in Sacramento or figuring out how to conduct personal finances in Tanzania. This was the opportunity Berners-Lee and his cohort of designers of the internet espoused—and it is the societal paradigm that Zuckerberg, Schmidt, and Bezos free rode, exploiting the open image and nature of the internet by projecting that their companies constituted a democratizing force, just as the internet itself does.

But the real consumer internet does not have such a profound or socially good purpose at its core. That is not to suggest that business leaders' remarks are necessarily insidious or disingenuous. But in this case, aligning his company's mission with the inspirational construct presented by Tim Berners-Lee supports Eric Schmidt's business interests. Google thereby fed off the public perception that the internet at large was a glorious gift to humankind—and it still does.

And now, like a stubborn weed snaking around the ankles of democracy, it has become the general view that internet companies promote society's good no matter the protocol or platform or business incentive in question. This is the active social paradigm proliferated and perpetuated by the internet barons. We, the public, have been duped.

Ecce Homos: The New Thomas Kuhns

Enter the Thomas Kuhns of the internet—the new thinkers working to dispel the virulent and misleading notion that Silicon Valley executives wish only to promote the public's interest. Since the 2016 presidential election, a series of bold ideas advanced by a cadre of emerging intellectuals has reshaped policy thinking concerning the internet.

There is Roger McNamee, the investor and adviser to senior technology executives, who asserts that he introduced Mark Zuckerberg and Sheryl Sandberg, Facebook's CEO and COO, respectively. McNamee has put forth the idea that despite his ongoing investments in firms throughout Silicon Valley, some of the leading internet properties—particularly Facebook—have imposed serious and systemic damage to democracies around the world.[1]

There is Shoshana Zuboff, the Harvard academic, who has eloquently written about how the emergence of surveillance capitalism has torn at the heels of major internet companies and who has depicted the capitalistically fluid nature of the companies' business.[2]

There is Tristan Harris, the former ethicist and designer at Google, who has cultivated the compelling narrative that because of the way social feeds on modern social media systems are designed, users are forced to sustain deep psychological harms. To relieve us of them, he argues, we must encourage companies like Facebook, Google, and Twitter to reorient their systems to ensure that time spent on their platforms is "time well spent."[3]

There are Zeynep Tufecki and Tim Wu, both brilliant professors and contributors to the *New York Times*, who have raised some of the most critical questions concerning the commercial nature of internet firms and their executive decisionmaking and have proposed that new policies must be developed to better protect competition and privacy, among other policy measures.[4]

And there is Barry Lynn, the competition policy expert who is head of the Open Markets Institute and formerly senior fellow at the New America Foundation, the influential Washington, D.C.–based public policy think tank that acrimoniously ousted him (and at which I have served as a fellow). Lynn, a thoughtful and pragmatic expert on market concentration and the public policy measures that can be taken to limit it, has for the past several years led a contingent of like-minded activists and intellectuals who have put forward a number of interesting ideas to diminish the monopolization of industries that the Open Markets Institute argues has plagued American consumers for so long.[5]

Over time, the public policy concerns around the businesses that operate over the internet have changed completely; the tone and tenor of the way we talk about the internet in the media has become increasingly incendiary. What was once considered a positive medium for sharing and collaboration, the public domain that could connect smart new ideas and people no matter who or where they might be, is now seen as a medium that bears countless social harms. From the spread of hateful messages that have led to mass killings to the coordination of ad campaigns that have prompted systemic and automated bias against marginalized classes, we have now seen it all—and thanks to academics and advocates such as Harris, Zuboff, Lynn, and countless others, the common conception among policymakers is that something must be done to improve the situation of internet consumers.

The ideas that have been presented by this striking band of thinkers have encouraged many others. Traditional internet and consumer advocates—hard-liners who for many years have advocated the need for progressive changes to the way we regulate the internet—have found a new ledge to stand on. Whereas in the past some might have found it difficult to argue that the internet is causing harm and that regulation is necessary, there is a newfound zeal and courage among the ranks of American advocates—particularly those that have in recent years maintained closer relations with the industry, such as the Center for Democracy and Technology and the Open Technology Institute. Even organizations that have purposefully been kept at more of an arm's length from the industry—such as the Open Markets Institute, Public Knowledge, the Electronic Privacy Information Center, and the Center for Digital Democracy—have found a new voice because of the stark resolution of anti-industry advocacy. Thanks to the Thomas Kuhns of the internet, we have entered new intellectual territory.

This new school of thought has furthermore encouraged a palpable revolt among the individuals who actually deliver value in Silicon Valley—developers and engineers. Hardly a day passes that we do not see a new perspective offered by a former employee of Facebook or Google who is willing to speak out about how the industry is actively encouraging societal harms. Sandy Parakilas might be the best example from recent years. Featured in a *60 Minutes* exposé of Facebook, Parakilas insinuated that, despite his personal outcry, the company actively encouraged privacy-invading practices by using personal data in seemingly insidious ways.[6] There are many more examples now. Chamath Palihapitiya, formerly a vice president at Facebook, commented in 2017 that the "short-term, dopamine-driven feedback loops that we have created are destroying how society works. No civil discourse, no cooperation, misinformation, mistruth. . . . This is not about Russian ads. This is a global problem. It is eroding the core foundations of how people behave by and between each other. . . . I can't control them. I can control my decision, which is that I don't use that shit."[7] Palihapitiya has since moderated some of these statements, but his original sentiments are clear.[8] Guillaume Chaslot, the former Google engineer who has designed a technical system that attempts to enable study of the YouTube recommendation algorithm, has noted that the system "isn't built to help you get what you want—it's built to get you addicted to YouTube."[9]

Beyond the many critics of the consumer internet business model are new groups of employees who feel increasingly comfortable in holding their employers accountable on fundamental social concerns. Google presents the most visible example: employee groups have led protests at the company on many counts, including the company's potential engagement with U.S. Customs and Border Protection, the federal agency within the U.S. Department of Homeland Security that is responsible for securing America's international borders. Some 600 Google employees petitioned, noting that they "refuse to be complicit. It is unconscionable that Google, or any other tech company, would support agencies engaged in caging and torturing vulnerable people." As Cat Zakrzewski noted in the *Washington Post*:

> The Trump era has sparked a Catch-22 for the company as criticism surges across the political spectrum. The search giant is trying to appease liberal employees who are increasingly taking their beef with the company's positions public, while simultaneously weathering accusations from Republicans—including the president—who say the company is politically biased against conservatives. . . . Yet taking a strong stance against working with the Trump administration's immigration agencies could strain an already tense relationship between Google and the Trump administration.[10]

All this outcry is forcing a much-needed impact on the public's perception of the consumer internet industry's overall contribution to American society. About that, there can be no doubt.

As British Prime Minister Harold Wilson once noted, a week is a long time in politics. Silicon Valley once stood untouchable. But the recent criticism of the consumer internet industry's business model has encouraged even national political leaders in the United States to issue sharp rhetoric—and in some cases back it up with action. Policymaker concerns regarding perceived overreaches of Silicon Valley start with the president himself. Whatever one might think of the genuineness of his intentions, President Donald Trump has seemingly plucked lines directly from Barry Lynn's Open Markets Institute team, noting that the likes of Google, Facebook, and Amazon present a "very antitrust situation."[11] He pulls at conservative heartstrings to formulate his argument, proclaiming that the algorithms developed and used by these companies fail to offer the user politically balanced perspectives when it comes to American politics or his presidency itself, as his Twitter posts demonstrate:

> Google search results for "Trump News" shows only the viewing/reporting of Fake News Media. In other words, they have it RIGGED, for me & others, so that almost all stories & news is BAD. Fake CNN is prominent. Republican/ Conservative & Fair Media is shut out. Illegal? 96% of results on "Trump News" are from National Left-Wing Media, very dangerous. Google & others are suppressing voices of Conservatives and hiding information and news that is good. They are controlling what we can & cannot see. This is a very serious situation—will be addressed![12]

> Something is happening with those groups of folks that are running Facebook and Google and Twitter, and I do think we have to get to the bottom of it. It's collusive, and it's very, very fair to say we have to do something about it.[13]

> Wow, Report Just Out! Google manipulated from 2.6 million to 16 million votes for Hillary Clinton in 2016 Election! This

> was put out by a Clinton supporter, not a Trump Supporter! Google should be sued. My victory was even bigger than thought! @JudicialWatch.[14]

Trump's concerns fueled a major inquiry by former U.S. attorney general Jeff Sessions, who issued a broad invitation to state attorneys general to convene with him and discuss the possibility of driving greater transparency into the ways in which the content-prioritization algorithms used by these companies are developed, to the end of determining whether the president's broader implications of bias against conservatives might actually be true.[15] To date, all three branches of the federal government have raised new antitrust inquiries, with investigations led by the Justice Department[16] and Federal Trade Commission,[17] the House Judiciary's Subcommittee on Antitrust, Commercial, and Administrative Law,[18] and an expansive group of state attorneys general.[19]

This is an illustrative bellwether of the kind of regulatory comeuppance we may see over the coming years. These inquiries and related inquiries are still under development, and whether any regulatory results will emerge remains unclear. The feelings underlying such efforts persist and have seeped into the heart of American politics—including the U.S. Senate. Senator Ted Cruz (R-TX) presaged Trump's inquiry, asking Mark Zuckerberg during his April 2018 congressional testimony whether Facebook represents "a First Amendment speaker expressing your views or . . . a neutral public forum allowing everyone to speak?" Senator Cruz added in the same hearing that "there are a great many Americans who I would say are deeply concerned that Facebook and other tech companies are engaged in a pervasive pattern of bias and political censorship."[20] The allegation that these companies have organized among themselves inordinate amounts of power, so much so that they can negatively influence the social welfare, was overtly made by Senator Lindsey Graham (R-SC), who asked Zuckerberg point blank whether he believed his company was a monopoly.[21]

The inquiries have stretched across the aisle; in fact, most Democrats would quite likely contend that it was they who initially fueled the ongoing techlash. Senator Mark Warner (D-VA) has been on record since the presidential election in 2016 about the need to force changes to the way internet commerce works and in October 2017 proposed the Honest Ads Act to the Senate.[22] Joined by Senator Amy Klobuchar (D-MN) and the late–Arizona Senator John McCain (a Republican who was later replaced in cosponsorship by Senator Graham), Warner has been the most vocal advocate for political ad transparency and consumer privacy, among other matters of technology and telecommunications regulation.[23] Senator Elizabeth Warren (D-MA), from the 2020 presidential campaign trail, expressed great concern about growing market concentration across various sectors[24]—very much in line with the perspective of the Open Markets Institute, which has outlined how dozens of industries have significantly increased in market concentration in recent years. Representative David Cicilline (D-RI), having shared novel legislative and regulatory ideas with his congressional colleagues through his leadership of the House Subcommittee on Antitrust, Commercial, and Administrative Law, has shown tremendous thought leadership and political courage in attempting to address the overreaches of the technology industry with action.[25] Senator Ron Wyden (D-OR), meanwhile, has suggested that Mark Zuckerberg should potentially face a prison term because he has "repeatedly lied to the American people about privacy" and "ought to be held personally accountable."[26]

This political rhetoric is not exclusive to the United States. Other jurisdictions have already taken far more strident steps. The European Union, long a bastion of individual privacy rights and the maintenance of the economic strength and intellectual independence of the individual, has laid out a number of innovative rules and regulations that will significantly impact internet commerce if they are upheld to the word. This starts with the EU's General Data Protection Regulation (GDPR), the novel regime pertaining to the collection and use of data associated with EU citizens that went into effect in May 2018.[27] The United Kingdom, on the heels of Brexit, followed suit with a scathing parliamentary committee report on Facebook and the disinformation problem, describing the actions of the firm and its chief executive as those of "digital gangsters" and suggesting that Zuckerberg will be held in contempt should he set foot in the United Kingdom as long as he fails to respond to Parliament's many inquiries about his company's actions.[28] The unprecedented joint International Grand Committee on Disinformation and Fake News, chaired by Canadian Member of Parliament Bob Zimmer and composed of high-ranking officials from many nations, has called for the senior leaders at Facebook—not its local staffers or lower-ranking policy officials—to show up and testify before the committee.[29] The United Nations Conference on Trade and Development convened expansive meetings in 2019, in reference to which Secretary-General Dr. Mukhisa Kituyi noted that "Digitalization . . . has led to winner-take-all dynamics in digital markets," and that "economies of scale and network effects have led to single dominant firms in e-commerce, online search, online advertising, and social networking," which has "given these firms significant control over consumer data."[30]

The Commission nationale de l'informatique et des libertés (CNIL), the French office responsible for data regulation, has conducted numerous investigations and levied fines against the industry, including a fine of €50 million against Google "for lack of transparency, inadequate information, and the lack of valid consent regarding the ads personalization."[31] Australia's commercial regulatory agency, the Australian Consumer and Competition Commission (ACCC), has suggested that the internet industry has visited tremendous harm on consumers; the agency has made rigorous new proposals to advance technology regulation.[32] Japan has opened discussions around new legislation to regulate digital giants, which are perceived in the country to have harmed market competition.[33] It has also set up committees to explore the potential impact of Facebook's Libra cryptocurrency on monetary policy and financial regulation.[34] Singapore and Germany have proposed stringent new content policy standards targeting hate speech disseminated over social media platforms.[35] In Belgium, a Brussels court ruled that Facebook had broken privacy laws and ordered the company to delete the illegal data, although Facebook has challenged those claims.[36] The Italian privacy authority (Garante per la protezione dei dati personali) has charged Facebook with misleading users and mishandling data.[37] The Dutch Data Protection Authority (Autoriteit Persoonsgegevens) fined Google in 2014 for breaches of privacy policy and since then has investigated the industry on other counts.[38] Brazil and Argentina have updated their privacy laws and regulatory regimes in recent years, putting the internet firms on watch.[39] The list goes on and on.

The inquiries have reached a local level in the United States, too. California has passed a much-anticipated privacy law, lauded as the most stringent in the country.[40] Illinois has attempted to enforce a powerful new biometric privacy law targeting Facebook's facial-recognition technology.[41] New York City has convened an expert group to examine how algorithmic transparency and interpretability are critical to maintaining fairness for residents (though questions have been raised about those efforts).[42] Hawaii, Maryland, Massachusetts, Mississippi, and many other states have either taken similar action or are expected to follow suit over the next few years.[43]

The commercial engine underlying Silicon Valley appears to be under attack.

The Consumer Internet: A New Paradigm

And so a new paradigm shift—riding the wave of the techlash—is in full swing. The world is descending on the consumer internet industry. People are angry, politicians are taking swings, and governments loom. The internet, once a forum for positive sharing, now carries the brand of promoting an industry of cutthroat capitalists who nip at the heels of American democracy. We have developed and adopted a powerful new theory of the case, just as Copernicus once did: The leading internet firms are not God's gift to the world. They can cause grave harm just as past industries have. At their heart of hearts, their nature is defined by profit-seeking in a manner that lacks any nonmarket-driven accountability to the public.

But this new condition raises a question: Given our sorry political circumstances, can anything really happen? Particularly if it is in the realm of polarized partisanship where economic regulation squarely sits? Will the full weight of our federal legislative and regulatory powers really be brought to bear on what is now the world's most profitable and powerful industry—especially in a period of gridlock in which the U.S. Senate cannot even execute a politically independent impeachment trial?

Consider the Zuckerberg hearings, as well as those over the past three years that have involved the other major consumer internet companies. Despite all of the public anticipation leading up to those congressional inquiries, all we got in the end were some memes of members fumbling as they confusedly questioned the industry. Recall, for instance, Senator Brian Schatz (D-HI), who asked a question about sending e-mails over WhatsApp, or Senator Orrin Hatch (R-UT), who asked about how Facebook makes money.[44] Since the hearings, these interactions have been explained away with rationalizations. Senator Schatz says that his was an earnest misstatement,[45] and Senator Hatch's office has fairly indicated that he simply meant to underscore earlier discussion from the hearing,[46] both of which are likely true but nevertheless have influenced the public attitude concerning Congress's ability to regulate the industry according to its economic merits.[47]

The underlying question remains: will these hearings and all of the accompanying congressional scrutiny over the industry have lasting impact? The first of these hearings, in late 2017, explored possible interference in the 2016 U.S. presidential election. Consider what new material was actually learned during its course. The lawyers representing Sean Edgett, Richard Salgado, and Colin Stretch—the legal executives at Twitter, Google, and Facebook, respectively[48]—effectively served their clients by not revealing anything that might encourage regulatory ardor and suggesting that their companies had simply been caught off guard by the Russians.

I continue to have faith in the political process—and as my colleagues Gene Kimmelman, Phil Verveer, and Tom Wheeler have suggested in the past, we need more congressional hearings.[49] The congressional forum and its lines of inquiry are the only way to force the industry to defend its practices—or accede to regulation. We need only to design more effectively the right hearings at the right time and in the process bring in the right people and ask the right questions—particularly about the nature of the companies' business practices and whether and how those practices tread on the American interest.

In the meantime, I would encourage a thorough intellectual reassessment of everything we think we know and understand about the causes of the harms against democracy that have been systematically perpetrated by this industry. Such an assessment must begin with the business model at the heart of the consumer internet.

Grounding an Analysis of the Modern Internet

Numerous industry officials, policy experts, and legal scholars have written about the business models of internet companies in varied contexts, but much of this analysis lacks depth. For instance, some have suggested that for internet firms, or more specifically social media companies, "the business model is targeted advertising." Indeed, these are the terms in which Zuckerberg describes how his company makes money, as he quipped in response to Orrin Hatch: "Senator, we run ads."[50] But the way Facebook truly operates is far more complex. Zuckerberg did not go into close enough detail to depict the business model in the resolution necessary for Congress to begin to address the internet industry's root problems.

Building on his judgment that the industry simply runs off ads, some have additionally argued that internet companies should be encouraged (or forced) to make their services available for some subscription fee as an alternative to their current business model premised on advertising. Yet if Facebook were to convert all of its "free" users into paying subscribers, relieving them of targeted advertisements and content curation in their social feeds, then perhaps we could effectively blunt the formation of polarized filter bubbles and diminish the disinformation problem.[51] But the idea that Facebook should simply switch to a subscription model to protect American democracy, as Roger McNamee, Jared Lanier, and others have suggested, carries deep flaws.[52] If subscription were a requirement for all users, the number of people using the platform would fall so drastically—especially in developing countries—that the benefits of social media would be severely diminished. Such wellsprings of free thought and expression over Facebook that prompted the Arab Spring would be stopped in their tracks, if all of the people who participated in the Arab Spring were suddenly required to pay $100 a year for Facebook access.

Even if users were given a choice between the continuation of targeted ads and content curation in their news feeds and having to pay a steady subscription fee, none of the problems that our democracy is currently facing would be earnestly addressed; they simply would be ignored. Imagine the entire American social media market being presented the opportunity to subscribe. How would users respond? At the rate of $100—or anything within that order of magnitude—not nearly enough people would switch. This leaves aside the question of users in developing nations that might feature greater political instability. Indeed, it is perhaps the people who would be most unwilling to make the payments for social media subscriptions whose news feeds we should worry most about. (Another option exists, too: a scheme whereby Facebook assesses your wealth—having inferred it through analysis of your personal information—and offers you a price-discriminating fee that by design is valued proportional to your spending ability. We can throw this option out the window, though. It remains highly unrealistic at this stage, as large-scale internet firms are not poised to take such a discriminating approach because of the obvious public outcry that would rise against the brand.)

Overall, replacing advertising revenue with subscription fees would be insufficient to meaningfully address the negative externalities perpetrated by Twitter, for example. This analysis leaves aside additional critical questions we would eventually need to answer. For instance, should subscription be a voluntary industry measure or regulated by the government? If the latter, should regulators impose the subscription restrictions only on large internet companies such as Twitter, or should they pull small start-ups into the regulation as well? If they choose not to regulate smaller companies, would users move to newer platforms that do not maintain the subscription requirement? And how much should companies charge for the subscription in the first place? Academics in the United States have estimated both the amount a typical consumer is willing to pay to use social media and a user's worth in annual dollar terms to the social media companies. But if Twitter were to charge the same flat rate for everyone, much larger proportions of people in developed economies would be able to pay for it, but not those in developing ones (unless the company pursued price discrimination programs)—effectively creating an imbalance in access. What would such a situation say of the world that we want to create?

A final concern about the subscription proposal: if all of the major companies were to enable the option due to regulatory requirement, how would consumers react? Each of the firms might potentially attempt to establish its own network effect given the new economic regime and as such might attempt to undercut rivals and attract users to its own platforms. Facebook, for example, might undercut prices set by Twitter, which might drive users off Twitter. It makes little sense to pay for a service that is more expensive, especially if Facebook can establish the one-to-many design through a new service that subsumes Twitter's business. Creating such a requirement might thus prompt further unwanted anticompetitive effects in the internet market. If consumers decide it only makes sense for them to pay subscription fees for one social media service, could there be deleterious effects for the consumer market overall? It is difficult to say.

Still others claim that the business of social media and the internet is all about data.[53] This, too, is an oversimplification of the problems at hand. Yes, the firms that sit at the center of the consumer internet collect inordinate amounts of data on the individual—more so than any other corporate entity or government in the history of humankind. Data collection must be a significant contributor to their businesses, otherwise they would not undertake it. But such collection of data has happened to varying degrees in other industries for decades. While the data-collection practices in other industries have never been as extensive as that of a company such as Google, the business model of Google cannot end there. How does it actually make money from those data? Why does it collect so much data? And why cannot other companies themselves make profits at the margins Google appreciates from such rich data collection? (Google was among the companies with the most cash in reserve at the height of the coronavirus outbreak in the United States.)

These are critical questions that need to be addressed on our way to defining the consumer internet business model and designing a remedy to contend with its overreaches.

The Contours of the Consumer Internet

Given the noise injected into the public's conceptualization of the consumer internet's business model, it is important to first define the contours of the consumer internet, which I believe democratic societies more broadly are most concerned with.

In recent years, these public concerns have become laser focused on a series of harms that have gravitated into the crosshairs of public outcry: the spread of hate speech, the disinformation problem, foreign election interference, algorithmic discrimination, terrorist recruitment, incitement to violence, and anticonservative bias. One could ask the pertinent question: Are these the harms that are indeed the most important to society when it comes to thinking about what the consumer internet has foisted on us? At one level, I think they absolutely are. They represent the uppermost layer—the externally observable symptoms—that is associated with a much deeper problem. They are the "here and now" that the people of Paris, Jakarta, Bombay, Christchurch, and Charleston care about today. They influence our national politics and the direction of our social countenance.

At a more critical level, however, is the business model of the consumer internet itself—the precise expression and manifestation of the commercial desires of the largest firms that occupy the center of the consumer internet splayed out against the theoretical economic and regulatory boundaries that the U.S. government has set for them—along with the commercial appendages and entities that the firms bring along in support of their core business, including the data brokers and ad exchanges. At the very center of the consumer internet—beyond the superficial manifestations of harm such as the disinformation problem or the spread of hate speech or the encouragement of persistent algorithmic bias—is a silent mechanism that works against the will of the very people whose attention, desires, and aspirations it systematically manipulates with cold technological precision.

This machine is principally responsible for the proliferation of society's concerns about the technology industry, including the terrible symptoms of disinformation and hateful conduct that we experience at the surface of the internet. And it is this machine that should be the subject of our policy analysis.

But how can we manage—or even begin to address—such a wide variety of problems engendered by a business model that is so staunchly defended by the robber barons of the internet behind the closed doors of Congress itself? We will need to incisively cut past the noisy exterior of the industry's advocacy about its self-proclaimed positive impacts on society and pry through the engineered spiderwebs into the heart of the problem. If we fail in this—if we fail to consider that the central business model of the consumer internet is so clearly responsible for all of these harms in the first place—then we will ultimately fail American consumers and citizens.

A Walk through Silicon Valley Today

What is the business model of the consumer internet? What do we mean by the "consumer internet"? And what separates it from other segments of the digital economy? Let us first examine a list of the biggest internet companies in the United States by revenue:[54]

	Revenue (US$ billions)	FY
Amazon	253.9	2018
Google	120.8	2018
Facebook	55.01	2018
Netflix^	15.8	2018
Booking^	12.7	2017
eBay^	10.75	2018
Salesforce*	10.5	2018
Expedia^	10.1	2017
Uber^	7.5	2017
Groupon^	2.8	2018
Twitter	2.44	2017
Airbnb^	1.7	2016
Workday*	1.56	2017

A number of observations can be drawn from this list. Missing are some major Silicon Valley firms that have been in the news recently—Apple among them. I have excluded companies such as Apple, Dell, Hewlett-Packard, and Intel because they are not primarily internet businesses. While elements of their businesses surely touch the internet—not to mention certain business practices they undertake that effectively impact consumers and citizens—the main portion of their revenues does not derive from operating services that run over the internet.

Apple's economic strength comes from its sale of electronic consumer-device technologies along with its hegemony over closed-source mobile-software technologies operated exclusively over its devices—including iOS, the App Store, and the bundled ties established between its many other services—that it has developed over the past twenty years. Some of these companies will be the subject of inspection in various respects in my analysis, but they do not constitute the central part of the internet economy; they did not inflict public harms such as the disinformation problem that have been of concern in recent years.

The list also excludes another set of well-known software companies—Microsoft, IBM, Oracle, and SAP among them. These companies, too, do not make the bulk of their revenue from operating internet services. Instead, they sell software, be it to consumers directly or, as is more often the case with these four firms, to other businesses. This combined focus on software and so-called B2B services disqualifies them from direct inspection here. Again, there is no doubt that some of them operate key consumer services over the open web, Microsoft's Bing search engine being perhaps the most visible example. But these services are as a general matter neither a core functionality nor a key contributor to the firm's global revenues and, more important, they are not the commercial engine of these companies. By corollary, these ancillary services do not "matter" as much to the respective company. Glancing at the angles of advocacy pursued by these firms suggests as much. Microsoft, for example, has long been a proponent of privacy regulation, far more so than Facebook or Google, since the company does not make its lion's share of profit off data and advertising.

Some of the companies listed above are primarily B2B businesses that simply operate over the internet. (These firms are marked with an asterisk.) They are apparently doing little harm to democracies overall, particularly since their interaction with individual users is minimal. Thus I exclude them, too, from the core analysis here.

Finally, of the remaining companies on the list, some interact with the individual user far less than others do. Airbnb, Expedia, and Uber fall into this category and are marked with a caret symbol in the list above. These companies might set cookies on your browser to infer where you live, what kind of internet connection you have, what sort of computer, mobile phone, and browser you are using, and what business or vacation destinations you search for over their platforms—all to the end of determining some measure of your propensity to spend money on their respective hotel, rental, travel, and housing services.

But they do not compare with the platforms operated by firms like Facebook, Google, and Twitter in the level of individual interaction with the consumer. In every sense, these three companies are a world apart; they are highly dialogical with consumers, in that they participate in intelligent back-and-forth engagements in real time. The consumer scrolls through feeds, repeatedly hovering over certain links, engaging with certain posted videos, clicking through to view certain news articles, watching ads, and interacting with friends, colleagues, acquaintances, journalists, thought leaders, celebrities, and politicians. Meanwhile, the platform collects information about the consumer's viewing habits or interactions. As much as Expedia might want to employ such practices to increase its ability to monetize the user's experience with the company, it simply cannot compete on this front with the likes of Facebook. It has less material with which to engage with the end consumer and less of a platform over which to foster sophisticated dialogue through content curation, so attempting to do so would carry it so far away from its primary business that the firm's core value proposition would suffer from the unwanted distraction of attempted individual engagement.

Something important links the remaining firms in the list—specifically Facebook, Google, and Twitter—to one another. Yes, they are more dialogical with the user. But why? It is primarily because they enable any individual user to upload digital content that other users can see—they are, in other words, social media firms. They have outsize capacity to infer the nature of the user's personality and to monetize that capability.

What of the other firm in that list, sitting above all the rest: Amazon? It is a complex business and, strictly speaking, a combination of various of the preceding company types. Historically, it has operated more like Airbnb, in that it has operated services over the internet but largely has not engaged with the user in dialogue as YouTube or Facebook do. That, however, is changing. In 2018, Amazon broke away from the pack of firms that were orders of magnitude behind Facebook and Google in digital advertising revenue. It now sits in a clear third place, with more than 7 percent of the market.[55] Reading between the lines, it must be that a corporate strategy of Amazon's in recent years has been to infer users' personal desires and preferences by showing the user appropriate ads—making the firm much like Facebook and Google in certain respects. On Amazon consumers engage with products and services that others (or Amazon itself) have posted, and in doing so consumers reveal details concerning their interests, which Amazon can then use to operate a robust ad-targeting platform. And while this is not the prime revenue stream for Amazon—a firm that, among other lines of business, operates Amazon Web Services, the most popular American cloud-service provider, ahead of Microsoft Azure and Google Cloud[56]—its ad-targeting platform represents a major presence in the consumer internet industry.

These are only my perspectives. Some might contend that I am incorrect about the boundaries of the "consumer internet industry" in certain respects. Others might suggest I have grouped various firms in the wrong way, that certain of them belong in another group, and so on. But there is an easy test of this: examine the policy advocacy objectives of the various firms, which can reveal a company's underlying profitmaking interests in explicit terms. And there is no better way to examine this sort of advocacy than to look at the chief executives' statements. American technology chief executives enjoy an admired and rarefied existence with the public, but when push comes to shove, they are also the company's first and foremost corporate advocate. On behalf of their shareholders, they have to be. Mark Zuckerberg, Sheryl Sandberg,

Top Five Companies, Ranked by U.S. Digital Ad Revenue Share, 2018 and 2019

% of total ad spending

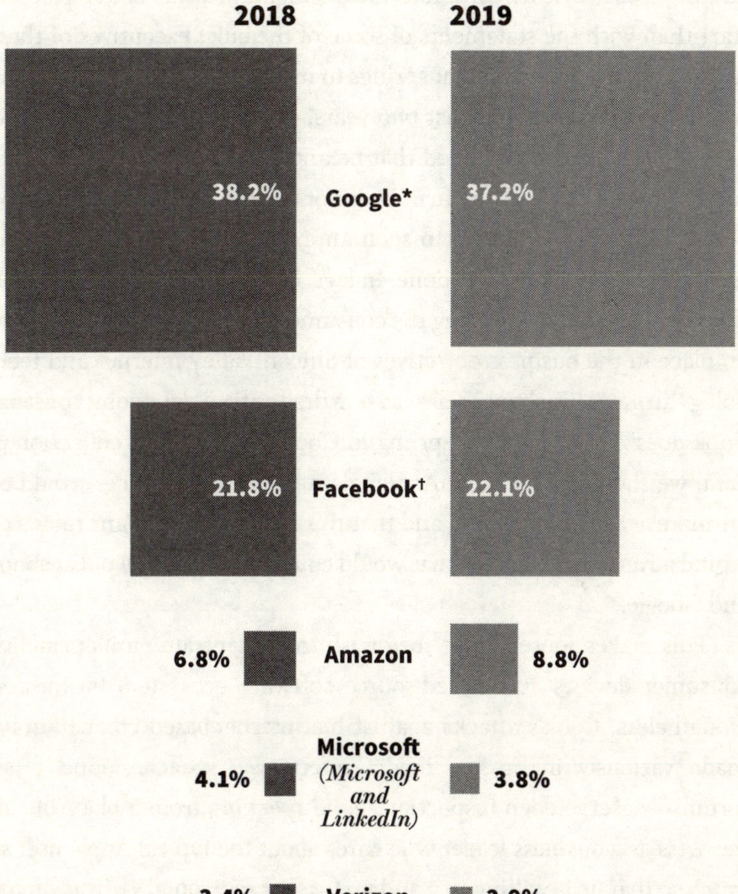

Note: U.S. total digital ad spending in 2019 = $129.34 billion; includes advertising that appears on desktop and laptop computers as well as mobile phones, tablets, and other internet-connected devices; includes all of the various formats of advertising on the platforms; net ad revenues after companies pay traffic acquisition costs to partner sites; *includes YouTube advertising revenues; †includes Instagram advertising revenues.

Source: eMarketer, February 2019.

Sundar Pichai, and Brad Smith—their voices project their companies' perspectives faster and further than any registered lobbyist ever could.[57]

It is useful to test the boundaries I have suggested thus far by examining some of this corporate messaging, and what better place to start than with the statements of some of the chief executives of these firms? Perhaps the name that springs to most people's minds as Exhibit A is Tim Cook. Over the past two years, Cook, the chief executive officer of Apple, has proclaimed that he and his company stand for consumer privacy and that the firm does not use the personal information pertaining to its customers to such an extent that some internet and technology companies have done. In fact, he is even more explicit about this: he notes that monetizing the consumer's personal data should have no place in the business objectives of Silicon Valley internet and technology firms. This should serve as our first critical data point: because Cook does not hesitate to reprimand Google's ubiquitous collection of data, we should expect that Apple's data-gathering practices would be far more acceptable to users and that it is not in Apple's plans to craft a digital advertising network that would compete at the level of Facebook and Google.[58]

This makes sense: Apple may wish to concentrate on its principal consumer devices and closed-source software ecosystem businesses. Nonetheless, Cook's attacks against his internet-based competitors—made variously in op-eds, media interviews, speeches, and other forums—deserve deep inspection. Is he speaking from a place of earnestness as a business leader who cares about the typical Apple user so much so that he is willing to stand up for that individual's privacy more than other corporate executives would? Or could it be that Cook has a distinct commercial incentive to make such statements?

I would strongly suggest the latter.

It is clearly not within Apple's business interest to monetize consumer data to the extent that companies like Facebook have done. While Apple does collect inordinate amounts of data—particularly through its customers' use of popular devices such iPhones, iMacs, laptops, and

Apple watches, not to mention the firm's proprietary software, including all of the Apple apps, such as Maps and Mail—Cook's point is that his firm does not use that information to make money.[59] Or, stated more precisely, Apple does not make money off this data by directing targeted ads to customers, although collecting the data itself is still in Apple's business interests to monitor and improve the company's product offerings. Given this, we can conclude that Apple is not a key member of the consumer internet industry.

Thus it is clear we cannot interpret Cook's withering critiques as some modern form of *noblesse oblige*; indeed, it is quite the opposite. What is less clear, though, is why Cook is attacking his Silicon Valley counterparts. Is it to stand up for the American consumer? Or is it to contend that his firm has a more compassionate heart than others? Why would his firm not use customer data for advertising if there were a commercial path to doing so at a profit? That is, after all, Apple's *raison d'être* as a matter of economics—simply put, to maximize profits in the long-term interests of the shareholders. The conclusion must be that he has decided that, all things considered—which could include current regulatory standards, additional forthcoming regulation, the ongoing techlash, the inability to innovate or compete, the opportunity to throw the competition under the bus, and the chance to demonize rival chief executives while highlighting his own sensitivity—it is actually in Apple's commercial interests to have its chief executive project the idea that Apple cares more about your personal privacy than Facebook and Google do, while leaving the immediate revenues to be had from targeted advertising on the table. Apple's braggadocio concerning its privacy standards benefits the Apple shareholder.

After all, when was the last time an American business leader stood up primarily for the benefit of the American people as a general matter? That is not a feature of our economic system; our country has cultivated an economic design over three centuries that favors the ingenuity of industry in such a manner that encourages industry leaders to be adventurous and daring, take investment risk, and develop the innovations that Americans will buy tomorrow. Under this regime, American society itself has suggested that business should not first think of regulatory boundaries but rather the theoretical economic space in which it can operate to be a vast, open space of opportunity. The first firm that can invade and occupy that vacuum can perhaps even achieve monopoly, at least for a time.

In effect, and taken with this economic environment in mind, that Apple has chosen not to pursue the monetization of certain forms of consumer data (at least, to date) should not be seen as a signal of its commitment to consumer rights. Apple has concertedly chosen not to use personal data simply because it is not in the company's best interests to do so. It is, in other words, a strategic decision. This is how the public should view these statements. They are nothing more, and nothing less.

If any doubts over this might linger, consider that Apple does not hesitate to compromise common conceptions of human rights should such oversteps be in the firm's favor to undertake. Take, for example, Apple's activities in China. In May 2017, the Chinese government issued a draconian data-security regulation, one that was particularly stringent against internet and technology firms, especially American ones.[60]

The much-anticipated regulation was loud and clear. A foreign company that serves Chinese nationals and wishes to collect and maintain data on Chinese citizens must engage in an extreme form of what is known as "data localization": the data must be stored on the Chinese national storage system in China. More questionably, foreign firms that maintain data on Chinese nationals must partner with a Chinese cloud-computing firm to build and maintain their Chinese data centers. Any data stored in China are subject to inspection by the Chinese government should it wish to investigate any matter, without exceptions. Finally, should the firm wish to transmit or transfer any data outside of China for any reason, the transmittal would be, per the regulation, subject to review and potential blockage by the Chinese security authorities. In a country that offers little transparency into the actions or motivations of a national government that has openly and consistently committed human rights violations, particularly against those thought leaders and activists who have tried to expose how the Chinese government has perpetrated these intense privacy violations against its own people, a data-security regulation such as this one should give pause to any multinational American firm—let alone a chief executive from Silicon Valley who preaches that other corporations should consider the position of the consumers' right to privacy.

Not so for Apple. Here was the ultimate chance to express resistance to the nature of the Chinese data-security regulation, but mere days after it went into effect, the firm disclosed—through a brief press release that failed to receive much scrutiny from the American media—that it had secured a deal to open a new data center in Guizhou in partnership with a Chinese cloud-security firm.[61] Apple quite likely decided, quickly, to comply with the Chinese regulation principally for three reasons. The first is that the firm wishes to protect its existing market share and customer base in China. China is a massive consumer market for Apple, perhaps its biggest projecting into the future, and if it were to choose to disregard the data-security regulations, the government could well choose to oust Apple from the market.[62] Second, Apple's primary manufacturing base is in China. It exploits the relatively inexpensive but talented Chinese labor pool available to its commercial advantage.[63] Third, Apple wishes to maintain a working relationship with the Chinese government so that it can maintain its Chinese market share and manufacturing base.[64]

With that cool calculation, Apple quietly subscribed to a Chinese regime that is never shy in silencing those who protest its policies and practices—whether through harassment, confinement, or torture.[65] Apple knows that very well and in complying with the regulation it basically shrugged and said to itself, "Well, that's fine." And it is likely—perhaps even inevitable—that some among Apple's ranks would have questioned the firm's decision given China's skullduggery. But these independent voices, if they exist, never became public. In this situation, then, all the public has to assess Apple's care for the world is the company's overt decisionmaking in the marketplace. Where, then, does that leave Tim Cook's statements that superficially juxtapose Apple's commitment to consumer privacy with the lack thereof with other companies? And what of the impression that he tries so hard to cultivate, that he and his company have attempted to cultivate around their emotive care about individual rights and global progress? Does it matter that Apple, at least in practice even if not in functional morality or philosophy, uses American customers' data for monetization less overtly than Facebook does but simultaneously submits to the Chinese government in the name of corporate growth?

I do not think it should. Cook's histrionics are so clearly designed for strategic reasons that I am quite surprised that the media continues to engage his calls for the regulation of social media (for instance, in the case of his calls for a federal privacy bill).[66] That is not to say that social media should not be regulated—it of course should be. But where social media firms should be regulated for their breaches of human rights, Apple, too, should be regulated—or at the very least adequately scrutinized by independent parties—for its disregard for the civil liberties of the 1.4 billion people of China.

To our earlier point, these facts should clarify why Apple does not qualify as a consumer internet firm, at least in the formulation offered here. Otherwise, Tim Cook would not without provocation assert that Facebook and Google should be regulated for the collection and use of consumer data. Additionally, I would suggest that as we are deeply inspective of Cook's false hubris, we similarly scrutinize the statements of other executives. One example is Marc Benioff, the founder and chief executive of Salesforce, who has suggested that "Facebook is the new cigarettes. . . . It's addictive. It's not good for you. There [are] people trying to get you to use it that even you don't understand what's going on. The government needs to step in. The government needs to really regulate what's happening."[67] Benioff is absolutely right in expressing concerns about how the use of social media triggers a response of addiction in many users, especially among children.[68] But like Apple,

Salesforce is not a consumer internet firm; it does not benefit from the monetization of consumer data to the extent that Facebook and Google do, given its core function of serving as an online-enterprise service platform for business clients.

Microsoft presents yet another example. The *New York Times* noted in a review of Brad Smith's book that the company "has positioned itself as the tech sector's leading advocate on public policy matters like protecting consumer privacy and establishing ethical guidelines for artificial intelligence." In response, professor David Yoffie of the Harvard Business School precisely summarizes the company's motive: "Microsoft can afford to be more self-righteous on some of those social issues because of its business model."[69] Smith's perspective that when "your technology changes the world, you bear a responsibility to help address the world that you have helped create," and furthermore that government needs "to move faster and start to catch up with the pace of technology," is motivated by strategic positioning; it aids Microsoft to attack a business model that it does not pursue but that Facebook and Google do. Microsoft did not voluntarily suggest it bears a responsibility to the public when it settled charges with the U.S. government after an antitrust investigation into the company's alleged anticompetitive practices.

It is not that Tim Cook, Brad Smith, and Marc Benioff make these statements aimed at the negative space of their respective firms' core business models principally because they feel the need to opine as public intellectuals. Rather, I would suggest that their issuing these statements—and, in the process, fueling the inspection of the true consumer internet firms—actually favors the long-term business interests of Apple, Microsoft, and Salesforce alike. This is true not only in the sense that their statements make these executives appear to care about economic equity and individual rights. In fact, unless they are stopped in some way, Google and Facebook will increasingly squeeze profit margins for most other major technology firms into the future. This is because Google and Facebook enjoy extraordinarily high profit margins yielding huge sums of cash—cash that they have smartly reinvested not

The Revenue Streams of the Leading Technology Firms Principally Determine the Statements Their Executives Make in the Media

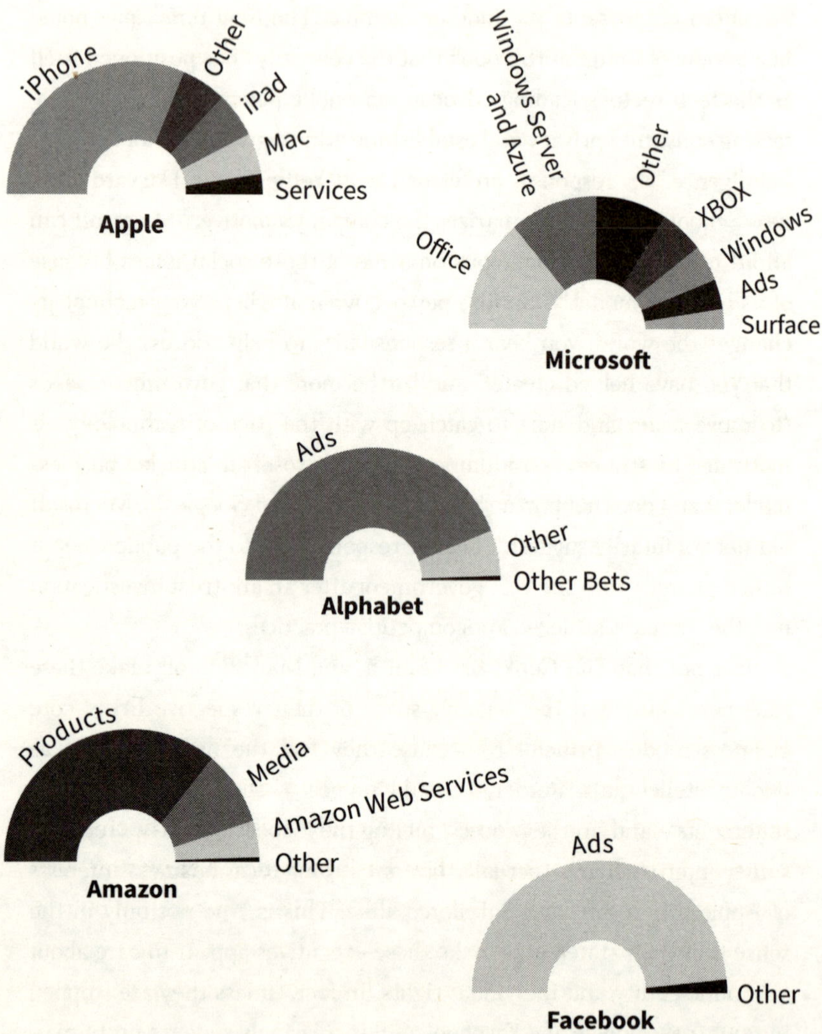

Source: Jeff Dunn, "The Tech Industry Is Dominated by 5 Big Companies—Here's How Each Makes Its Money," *Business Insider*, May 26, 2017.

only in their untouchable advertising regimes but also in younger side businesses that will increasingly challenge the likes of Apple, Microsoft, and Salesforce. In fact, the more that Facebook and Google profit from the consumer internet business model that they exhibit more than any other firm, the more they can invest in the development of ancillary products that will potentially subvert the other firms' leadership in their subsectors. Facebook's Workplace service, for example, is a major new investment for the company in the enterprise software market that Salesforce currently leads, and it will probably begin to threaten Salesforce's competitive edge. Google is making similar investments, and its foray over the past ten years into the consumer-device technology space—including cellular phones and home gadgets—may increasingly threaten Apple. And Google Docs, Google Sheets, and Google Slides directly challenge Microsoft's hegemony in enterprise software. This dynamic reinforces the idea that these various firms occupy different industries—and that firms like Apple, Microsoft, and Salesforce should not be considered consumer internet firms in the way that Facebook and Google are.

A New Perspective on the Economic Logic behind the Open Web

With a clearer picture of the firms that sit at the center of the consumer internet—including Facebook, Google, Twitter, and Amazon, in addition to second-order companies such as Pinterest and Snapchat—we can now analyze what makes their businesses tick. Let us start with the company that has been in the news the most—the one that puts up the big blue app, whose founder Mark Zuckerberg and Chief Operating Officer Sheryl Sandberg have been forced onto American television screens over the past three years.

Facebook makes the vast majority of its revenue from digital advertising—but if we were to stop there, we would be addressing just the superficial and leaving aside the full truth behind the company's operation. The Silicon Valley business model directs Facebook to create compelling services, harvest personal data used to create consumer profiles, and develop algorithms for content curation and ad targeting. These three pillars together define the vaunted institution that everyone in the digital industry venerates: Facebook's commercial engine. The first of these pillars, designing captivating apps that effectively engage consumers and win their attention over other competing services—is simple to understand but difficult to achieve. In fact, some economic experts might contend that only a big player in the consumer internet—perhaps one of the few biggest platform companies in the sector—can possess such engaging services.[70]

Consider Facebook's key offerings: the news feed that users scroll through to see a curated and ranked social feed of updates from across their network, the Messenger internet-based text-messaging service, the WhatsApp encrypted text-messaging service, and the picture-sharing service Instagram. These are the core services that are principally responsible for Facebook's perpetual growth around the world. It is difficult to imagine another competing service surviving in any of these market siloes under our current regulatory regime. Facebook's services have each claimed their respective markets by moving into them aggressively and establishing an extraordinarily powerful network effect that limits the capacity for would-be rivals to compete. Facebook's services benefit by raising barriers to entry; the competition's digital and physical infrastructures—extravagant capital expenditures—would have to be built up from scratch to reach the level of sophistication required to compete with Facebook. The highly vertically integrated nature of Facebook's business practices today—driven by the company's focus on the sequence of tracking user data, targeting ads at users, disseminating those ads over monopoly platforms, owning much of the infrastructure that enables those platforms to sustain themselves, and maintaining strong commercial relationships with vendors up and down the technology stack—represent a Goliath that not even the most enterprising David could overthrow.

Consumer internet firms have a stranglehold on the market for attention because of how engaging they are. YouTube's video recommendation algorithm, for instance, is engineered to capture users' attention and keep them watching—and inject a few targeted ads that might engage them as well. There is no other goal for YouTube; the only incentive is maximizing ad revenue. This is no public square. It is the dominion of a profit-seeking firm. And this is precisely where things get messy for the company: there are some especially violent types of videos—shootings, bombings, terrorist speech, exploitation—that many of the platform's users find highly engaging. YouTube does not know any better than to simply up-rank such videos and attempt to capitalize on the expected engagement they generate. Despite the unmatched commercial power of its parent (Google), and despite its powerful fleet of private servers around the world, YouTube simply does not have the incentive to deal earnestly with the problems it has created. And without appropriate legal exposure, it never will.

Tristan Harris has described the manner in which these firms capitalize on biological weaknesses in the human psychology, exploiting the dopamine-delivering effects that the platforms have on our thinking about the world. Harris and his colleagues appear to be correct in their analysis: these services are designed to be addictive. This propensity to addict the consumer to social media can cause great harm.

The second pillar—the collection of data to the end of creating and maintaining a behavioral profile on the individual user—feeds off the first pillar's effectiveness in engaging the individual. Consumers passively generate a great deal of exhaust in the form of personal behavioral data as they use internet platforms such as YouTube. Time spent viewing a given video, areas over which a user might have hovered the cursor or tapped the screen, the time of day and device type, the user's location, the browser and internet connection used to communicate with YouTube—all of these data are readily collected by the platforms to understand who users are and to target ads to them.

Without the collection of these data, such personalization is impossible. This marks a noteworthy distinction between the internet and media formats of the past. The ability for the most important modern media corporations—internet platform companies—to define the content users will personally see, the amalgamation of which is increasingly unique to the user over time, is a departure from the broadcast, television, radio, and initial internet communication regimes of the past. Harvard professor Cass Sunstein notes that such personalization can corrode the democratic process. The inability of an individual to have a clear conception of what the broader public consumes over internet platforms—unlike consumption of the more traditional media formats of broadcast television or radio—redefines the interaction between citizens and society in potentially harmful ways that have not yet been fully addressed or even witnessed.

The third pillar—the development of algorithms that curate our social feeds and target us with ads—feeds off the second pillar and fuels the first, completing the vicious feedback loop that has lit a fire at the feet of the American democracy itself. As data are collected on us, they are used in highly sophisticated machine-learning algorithms to understand who we are over time and algorithmically rank our social feeds in the manner these firms believe can maximize our engagement.

That, in short, is how the consumer internet makes money.[71]

The Subject of Our Analysis

Facebook, Google, and Twitter have each acknowledged that foreign agents did indeed engage in activities during the 2016 presidential election cycle that constituted nefarious "infiltration" of their platforms.[72] Throughout the subsequent official inquiry on this matter, the public has suffered one shellshock after another from a series of staggered revelations as to how Russian disinformation operators developed and disseminated politically charged content on the key internet platforms owned and operated by the three companies. For its part, Facebook has disclosed that up to 126 million users may have been targeted with

Russian disinformation.[73] The drumbeat of bad news appears so steady that the expectation now is that it will never cease to flow from the industry.[74]

The Russians' intent was most likely to interfere in the American electoral process and help elect Donald Trump to serve in the nation's highest office—and in the process, perpetuate chaos in the American political sphere. President Trump's win was accordingly celebrated throughout the halls of power in Moscow not because it represented better potential for strong diplomatic relations with Vladimir Putin but rather because Russian leaders believed his takeover of the Oval Office represented a bleak future for Moscow's sworn Cold War enemy—or at the least, a bleaker future than Hillary Clinton could have ensured.[75]

For Senator Dianne Feinstein (D-CA), the Russian activity constituted "cyber-warfare"—suggesting, she rightly believes, that the United States government must do whatever it can to defend the American people from the nefarious actions of foreign enemies.[76] For quite some time after the 2016 election, Americans pondered how our media ecosystem—the very medium to which our society had over time grown so addictively attached—could have been responsible for so much harm to the principles underpinning our democracy. Only a smattering of corporate forensics, industry analysis, academic research, and public inquiry could uncover what really happened: the Russians had figured out a way to infiltrate our beloved internet platforms and shower us with fake news to such a degree that many Americans would go on to manifest their collective, misinformed psyche inside the ballot booth, voting for the man who the Kremlin was desperate to see in power.[77]

But despite the intense scrutiny of internet firms by policymakers and the American people alike, few have implicated the business model underlying the consumer internet itself. This makes sense; it is a nonintuitive conclusion to suggest that the leading internet companies' business models themselves could have in part been responsible for promoting such insidious behavior, particularly given the corporate hoodwinking that industry executives have achieved in the face of national

Instances of Russian Disinformation Operations in the Lead-Up to the 2016 Presidential Election

Posted on: LGBT United group on Facebook

Created: March 2016

Targeted: People ages 18 to 65+ in the United States who like "LGBT United"

Results: 848 impressions, 54 clicks

Ad spend: 111.49 rubles (US$1.92)

Posted on: Instagram

Created: April 2016

Targeted: People ages 13 to 65+ who are interested in the Tea Party or Donald Trump

Results: 108,433 impressions, 857 clicks

Ad spend: 17,306 rubles (US$297)

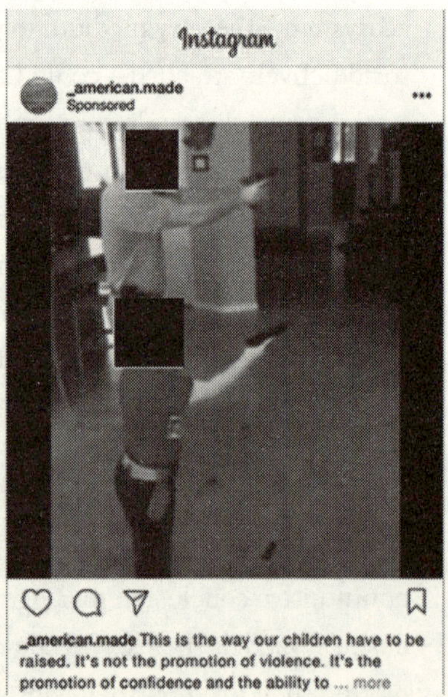

Source: Social Media Advertisements, Permanent Select Committee on Intelligence, U.S. House of Representatives, November 2017.

Posted on: Facebook

Created: October 2016

Targeted: People ages 18 to 65+ interested in Christianity, Jesus, God, Ron Paul, and media personalities such as Laura Ingraham, Rush Limbaugh, Bill O'Reilly, and Mike Savage, among other topics

Results: 71 impressions, 14 clicks

Ad spend: 64 rubles (US$1.10)

Posted on: Instagram

Created: August 2016

Targeted: People ages 18 to 65+ interested in military veterans, including those who fought in Iraq, Afghanistan, and Vietnam

Results: 17,654 impressions, 517 clicks

Ad spend: 3,083 rubles (US$53)

policymakers and the broader public. After all, the internet was meant to be an incredible democratizing force, not a new vector for subversion of good and protective order—and that is a brand that industry executives have readily appropriated for their commercial purposes. I believe, however, that it is time to scrutinize the industry much more harshly.

It may be that this depiction of the business model at play in the consumer internet is to some extent a simplification. If so, I suggest that it is not an unfair one. We cannot doubt that there are numerous appendages, exceptions, and parallel businesses that exist within consumer internet firms. For instance, Google might argue that its parent company Alphabet earns revenue from Google Cloud, Google Nest, and Google Assistant. Facebook might argue that much of its growth over the next ten years will center on new investments like the Oculus business, which does not necessarily align with the economic logic described here. And each of the companies that are the subject of the analysis in the following chapters might suggest that they are not the sole culprits engaging in the uninhibited collection of personal information or the acquisition of would-be rival internet media properties. All of this may well be true. But these ancillary issues are not the matter of central concern because they are not critical to the core issue that is at hand: it is the central business model of the consumer internet that is chiefly responsible for the harms perpetrated against democracies around the world. We must be laser focused on this troubling condition. We cannot allow extraneous information to muddy our independent perspective.

We need to revolutionize the regulatory regime that sits behind the consumer internet itself. Without doing so we cannot begin to address the harms perpetrated by the internet. My purpose in the following chapters is to comprehensively explore the three pillars of the consumer internet's business model to inform and ultimately to help determine what our democracy must do to combat the overreaches of the business model at the heart of the internet.

TWO

Data

The Harvesting of All Knowledge

The consumer internet industry collects immeasurable amounts of data on individuals—riches in personal information so vast that a single company such as Facebook might have millions of discrete, raw elements of data about users' behaviors, routines, and interactions over the internet. These data-collection practices have a clear commercial value; collecting the consumer's personal information feeds the business model that underlies the value proposition of the consumer internet to its users. The firms want to know about us, to the extent that they can exploit our individuality to the sole end of profit maximization. The tension between internet companies' keen desire to ingurgitate highly sensitive, personal information on us and our personal desire to maintain some semblance of privacy and autonomy has triggered a momentous, global, and public discussion over how best to shield consumers from the overreaches of unbridled, instinctive, and audacious internet commercialism.

These debates are ripe in many jurisdictions around the world. Europe, after years of consultation with a wide range of public stakeholders, including industry representatives and consumer advocates, introduced its General Data Protection Regulation in May 2018—a set of binding and stringent regulations concerning control of European citizens' personal information. France, Germany, the Netherlands, Belgium, and Spain have all taken active steps to manifest the broader European regulations into local national concepts to protect their consumer markets from economic exploitation at the hands of internet companies.[1] The United Kingdom has launched new investigations and deliberations into how to treat the commerce of personal data markets, particularly in light of the impact they have recently experienced from the disinformation problem and the spread of online extremism and terrorist activity.[2]

In Latin America, Brazil and Argentina have made progress in holding the industry more accountable with respect to its collection and use of consumer data.[3] Japan and South Korea have taken similar action.[4] Australia and Canada have also responded to protect their citizens, and India has followed with initial inquiries of its own.[5] And China and Russia have issued extraordinarily harsh regulations concerning data localization and security regulations that have forced the hands of American technology firms, many of which have yielded to the authoritarian governments and their demands of the industry, which are widely perceived as threatening human rights.[6]

These events have not gone unnoticed in Washington. Both the Trump and Obama administrations have engaged in formal inquiries to assess the commercial use of private data in the United States.[7] Additionally, Congress has long investigated matters of consumer privacy, including significant conversations that have taken place since the 2018 revelations of Cambridge Analytica's involvement in the 2016 presidential election.[8]

The time appears ripe to take advantage of the political will at hand and develop radical new thinking around a modern approach to mitigating once and for all the commercial data privacy problem.

The Modern Concept of Privacy

What do we mean by "privacy of the individual," and how should we think of it in the major policy debates we are having today? Privacy means different things to different people. For kids, privacy might mean assurance that embarrassing rumors are not spread to friends and classmates. For our parents and grandparents, privacy might have more to do with protecting themselves against the breach of personal identity information to maintain security of financial accounts. For workers, privacy might relate to their right to convene with colleagues and talk about their situation at work, perhaps even to unionize. Students and their parents might pursue educational privacy to bar their teachers, institutions, and guidance counselors from sharing some of their educational data with universities or potential employers. This is particularly critical in light of online test proctors who, during the course of the coronavirus outbreak, would, according to a *Washington Post* report, watch "students' faces, listen to them talk and . . . demand they aim their cameras around the room to prove their honesty."[9]

One of the early conceptualizations of the individual right to privacy came from a law review article published in 1890 by the American jurists Samuel Warren and Louis Brandeis, who noted that the right to privacy centered around the right to be left alone.[10] This conceptualization sparked a public conversation that continues today over the meaning of privacy as a human right worthy of protection.

With the rise of the information age, scholars and governments realized the need for a more robust framework that could memorialize institutional concepts and translate them into policies that industries and government should follow with respect to maintaining individual privacy commitments. These inquiries led to the development of the Code of Fair Information Practices, which was published in 1973 by the Department of Health, Education, and Welfare's Advisory Committee on Automated Data Systems. The department recognized the dire need for governmental guidance in contending with such rising novelties as computing technologies that increasingly intrude on individual privacy. The code advanced five governing principles to ensure individual privacy protection that would hold up as a powerful regulatory framework even today:

Notice and Disclosure. Organizations should not engage in any collection of personal information without full disclosure to the individual. Notice should also be given if the data collection might be used in making decisions that affect them directly—regarding, for instance, a loan or job application.

Individual Access. Organizations should not only disclose their practices to individuals but should also afford them the opportunity to access any information that the organization holds on them. Furthermore, organizations should be compelled to disclose how the individual's information is being used. In the consumer internet context, this might mean that Google should explicitly tell you that it is collecting your information from various sources and using it all to curate the content you see across the media ecosystem.

Purpose Specification. Organizations should not be allowed to collect information from the individual in one context and then use it in a completely different context. In other words, when consumers enter into agreements to share their information with a company, there should be some commitment from the company that the information will not be used in any other manner than what was originally agreed to. Many scholars have argued in recent years that there has been a gradual degradation in the industry's commitment to purpose specification, particularly in the context of the consumer internet.

Redress. In its most basic form, redress allows consumers to interface with companies and actively correct any incorrect data. As companies collect information on us, they might assume it to be accurate and true to our expressions in the real world—but this is not always the case. Much of the data that companies have on us is inaccurate. Our heights and weights might change over time, and we might have an interest in keeping that information updated in the books of our doctors and health insurance companies, for example, but not for other purposes. In the consumer internet context, Facebook might make hundreds of inferences about our personal behaviors and interests, including that we like a certain kind of music or a particular actor or genre of movie, but it might be that most of those inferences are no longer true—or never were.

Reliability and Security. Information collected for commercial use should be protected from any misuse, whether by the data collector itself or by a third party. In modern times, many have extended this principle to declare that companies should maintain robust security practices to protect their data from cybersecurity hacks.

This principled approach was notionally applied to the burgeoning economic activity over the consumer internet in the years after its birth and commercialization. Eventually, the code was translated into the Fair Information Practice Principles by the Federal Trade Commission, which issued guidance on data privacy and security to the industry.[11]

The Federal Trade Commission lacks the power to back up the guidelines effectively, however. At best, the principles are basic guidelines for commercial actors; by no means do they constitute regulation backed up by governmental enforcement. This is the reason for the stunning disparity between the guidelines and the consumer internet industry's actual practices. Companies like Facebook and Google simply do not follow the framework.

Data-storage capacity and speed of processing have advanced so swiftly that our regulatory framework has not kept up with the capitalistic extensions of the industry. As patrons of the consumer internet, we are left to fend for ourselves in the face of the Silicon Valley behemoths. It is a situation that breeds systematic economic exploitation.

We have left the information age and are now in the algorithmic age. It is an entirely novel commercial landscape, in which the business model dictates that the value of our information is not the details themselves but rather the amalgamation, inference development, and automated algorithmic application of the information. Layered under this new medium of commerce is the presumption we traditionally make in the United States to let the market roam free. But it is precisely this free market approach to regulating internet commerce that has led to the exploitation of the individual consumer. It has fomented a capitalistic urge for profit maximization at the expense of individual autonomy.

The Sensitivity of Fine-Grained Data

Imagine that corporation X collects minute details pertaining to your consumption behaviors. Maybe it is information about the purchases you have made on Amazon, or your history of search queries on Google. How are these an invasion of privacy?

Let us take an example that nearly everyone in the United States has to deal with every month: electricity consumption. You might imagine that this is a totally benign vector of longitudinal data about you; after all, who cares that you are using some amount of power every month? What is the worst that can happen—an environmentalist neighbor sees your monthly consumption levels by accidentally opening your bill from the mail?

Electricity consumption can be tracked at a level of granularity far greater than we might think on first glance. Long ago, after the commercialization of electricity, scholars of the electricity industry came to a determination that if the level of consumer demand across the network could be reasonably predicted, the power grid operator could route electric power across the network in an optimal manner. The invention of network analyzers allowed engineers to test electric power flows on a small scale through laboratory experiments, and in later years, computing software aided in the calculation of optimal power flows.[12] But a few years ago, the goal of accurately measuring granularized and real-time consumer demand—down to the household level—finally came to fruition as electric-power utilities designed and implemented advanced metering systems known as smart meters, which are currently in use in many regions throughout the United States and other parts of the world.[13]

These machines would replace the older meter models and were designed to report back to the grid your level of consumption on a temporal level. The idea was simple: the price of electricity can vary throughout the day; perhaps it is lowest at the dead of night and highest right in the middle of the working day, around two or three in the afternoon. This natural price variation occurs for a combination of reasons—but three of these are most significant: some power generation plants have to either be warmed up or be running at full capacity all of the time; the demand for electricity varies according to the number of people hooked up to the grid; and the industry lacks efficient ways to store excess supplies of power.

To mitigate the economic inefficiencies promoted by the natural world, we designed auction-based markets to invite concealed bids for aggregated supply and demand of power.[14] If the markets could know your historical minute-by-minute level of power consumption—and perhaps even know your real-time power-usage level—they could route power more efficiently by offering you electricity at real-time prices. Smart consumers might even choose to run certain high-energy demand functions—such as charging a smart car or running the washing machine—late at night to take advantage of low electricity prices. The most advanced meters—as opposed to the older models, which just cumulatively recorded your consumption over time so that the local utility could come by and check it once a month and apportion your billing accordingly—could know the real-time price of electricity in your local market, know your consumption preferences, orchestrate electricity consumption within your household according to those predetermined preferences, and automatically communicate back to the grid your temporally accurate consumption levels for both billing and routing efficiency purposes. To complete the loop, then, the most advanced meters typically record your fine-grained temporally precise power-consumption data and report it back to the grid. This level of syndication is indeed the direction in which the entire industry is moving.

This is a fascinating technology—it enables real-time markets and pricing, reducing economic friction in the process. For instance, advanced metering infrastructures can make it easier for you to sell your excess electricity or capacity—perhaps generated by your solar-panel installation or a rooftop wind farm or a Tesla battery—to other electricity consumers. Maybe even the environmentalist who lives next door. The figures on the following pages from researchers Mikhail Lisovich and Stephen Wicker of Cornell University illustrate what that fine-grained data might look like.

The first figure shows second-by-second electricity consumption levels in a typical apartment in upstate New York. The curious spikes might indicate certain kinds of activity taking place within the apartment. But Lisovich and Wicker applied a quite interesting twist to their experiments: in addition to hooking up a sensor to a student's breaker panel, they put a video camera in the apartment so that they could compare the video logs of what individuals in the apartment were doing—watching TV, taking a shower, and so on—with how much electricity they were consuming, according to the information gathered through the breaker panel. Then they worked backward and designed a prediction system that could help them infer the meaning of certain activities associated with spikes in electricity consumption as gathered through the breaker panel data, based on what they witnessed through the camera. In other words, they mapped typical movements in the power consumption data with actual activities in the household. The second figure shows what they came up with.[15]

Lisovich and Wicker were able to infer use of the microwave, the refrigerator cycle, the use of light switches around the apartment, and so on. A disturbing reality emerged: any person with access to your fine-grained power consumption data can use it to predict what you

Fine-Grained Consumption:
Aggregate Power-Consumption Data

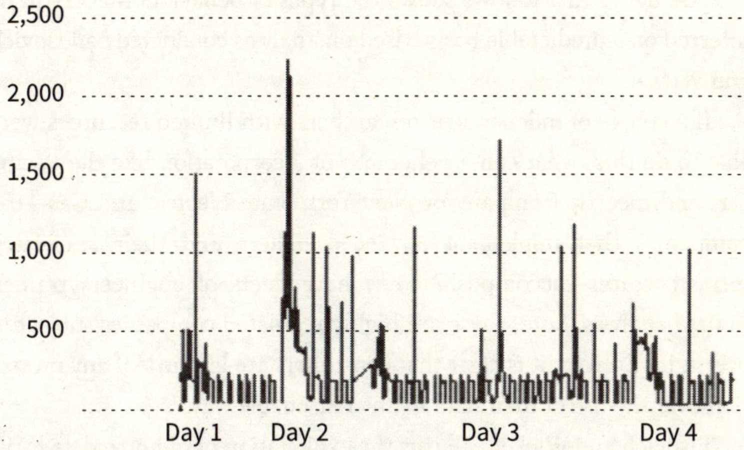

Algorithmically Inferred Behaviors:
The Derived Switch Events

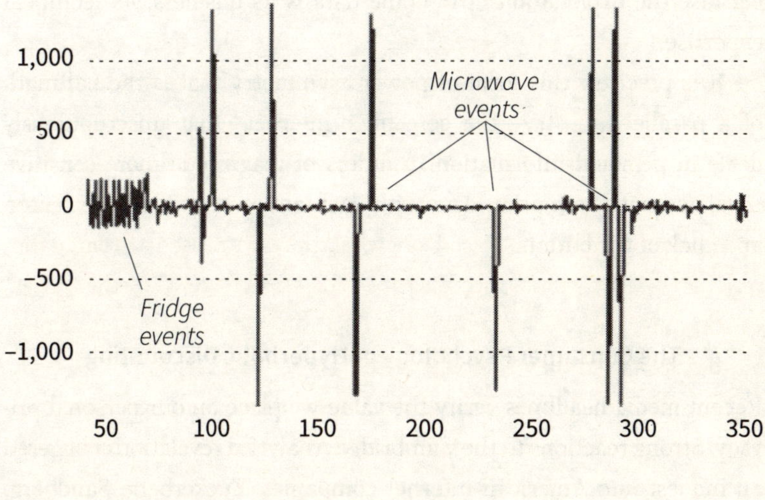

do at home and when you do it. This is highly sensitive information—the type of information that the typical American consumer would not wish to disclose in such high resolution even to close friends, let alone some unknown third party.

The figure that follows shows the types of behaviors that could be inferred on a predictable basis based on analysis conducted by Lisovich and Wicker.

If a couple of independent researchers with limited resources were able to do this, what can a behemoth of a corporation like the Pacific Gas and Electric Company or New York State Electric and Gas—the companies that implement advanced meters and their associated infrastructures—accomplish? They have fleets of engineers trained in data analysis. Sure, these are highly regulated companies subject to oversight. Even so, it is clear that consumers are left little if any protection over the use of their data in any reasonable way.

Lisovich and Wicker call this the exploitation of in-home data.[16] We cannot see it, we cannot comprehend its collection, and we cannot feel the economic value that is escaping our grasp and being deposited in the corporation's bank with every second. But the corporation knows because the exploitation of in-home data is its business, its technical expertise.

It is precisely this form of power asymmetry that is the hallmark of a parallel industry—the consumer internet—that unscrupulously deals in personal information in orders of magnitude more sensitive than power consumption data with each and every tap of the screen and click of the button.

The Consumer Psychology of Hyperbolic Discounting

Recent media headlines verify the value we place on our personal privacy. Strong reactions to the Cambridge Analytica revelations triggered an inquest into American internet companies. Zuckerberg, Sandberg, Pichai, and Dorsey as well as their lieutenants have all testified in

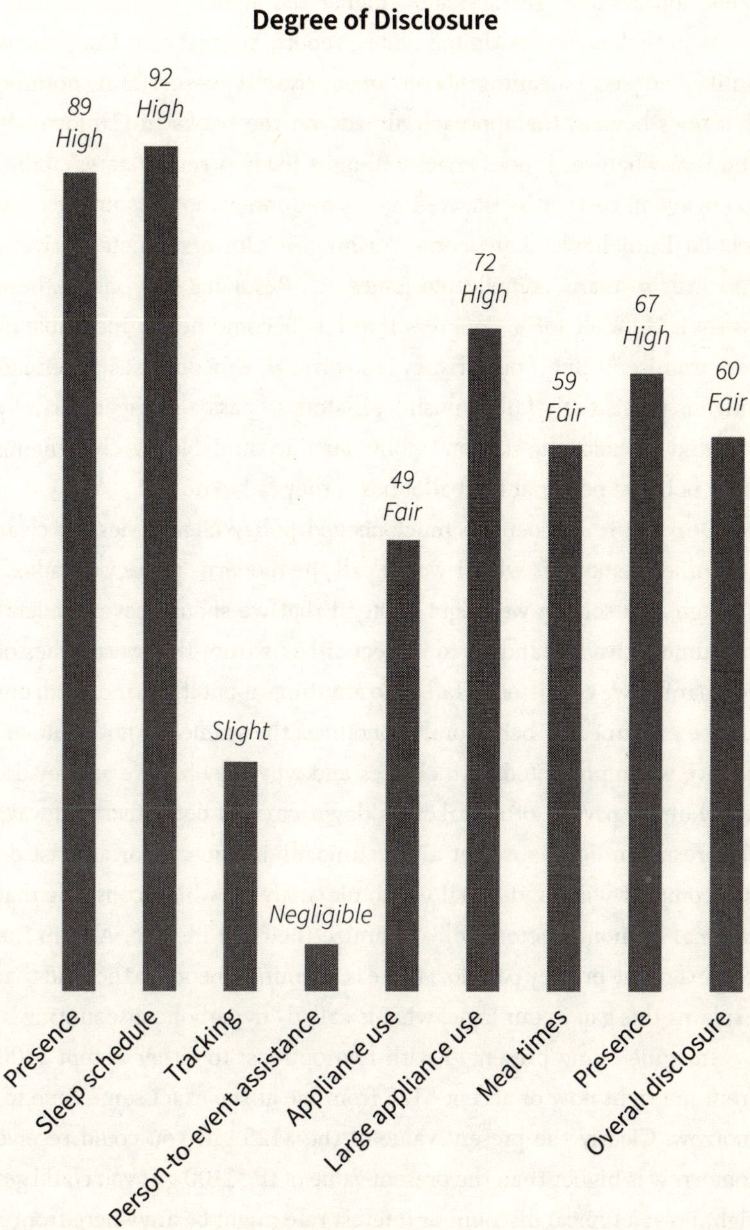
Degree of Disclosure
89
High
92
High
Slight
Negligible
49
Fair
72
High
59
Fair
67
High
60
Fair
Presence
Sleep schedule
Tracking
Person-to-event assistance
Appliance use
Large appliance use
Meal times
Presence
Overall disclosure

Washington and before governments throughout the world. And the tide of regulatory action, especially originating from foreign jurisdictions, appears to be rising steadily higher and higher.

Nonetheless, in the United States, reports suggest that Congress is unlikely to pass a meaningful consumer privacy law—certainly nothing that goes beyond the approach already on the books in California—and that whatever it does enact will most likely preempt states' ability to engage in further legislative inquiry and innovation. As privacy law scholar Daniel Solove of George Washington University notes, privacy "involves so many complicated issues . . . Resolving just one of these issues is difficult for a Congress that has become nearly incapable of compromise."[17] But if our privacy is so crucial, why do we as American citizens persistently fail to push legislators to pass a stringent privacy law and to hold our national politicians accountable by challenging their political power at the ballot box if they fail to do so?

Our failure to coerce a much-needed policy change despite clear consumer demand is what I would call the modern "privacy paradox." Though as a society we might contend that we should have a federal consumer-privacy standard to protect citizens from the overreaches of companies, we consistently fail to do anything about it. The conundrum can be explained by behavioral economics, the science of how humans behave when presented with choices and why they behave as they do. The human psyche often takes us down curious deductive pathways that result in decisions that are technically irrational—or at least do not comport with traditional economic analysis, which contends that rational economic actors will maximize their net income. And in the context of the privacy paradox, there is a seminal theory in the field that explains this gap in our logic, what is called "hyperbolic discounting."

Imagine being presented with two options: to either accept $100 from me right now or accept $125 from me at the exact same time tomorrow. Clearly the present value of the $125 gift you could receive tomorrow is higher than the present value of the $100 gift you could get right now; a typical discount or interest rate might be anywhere from 3 to 20 percent a year depending on who is loaning you the money, and at such rates, $100 would only grow by a few cents over the course of a single day.

Yet most people would probably opt to take the $100 now.[18] The principal conclusion economists have drawn is that, in general, valuation of the gift that offers immediate gratification is higher than its present value. In other words, people irrationally place a higher value on equivalent interest-adjusted payoffs that come sooner. This makes sense: if you get the $100 right now, you can use it right away. You don't have to wait an extra day, idling about and thinking over how you would use the (admittedly larger) sum in the future—and who knows whether you will need or want the money as badly then. By taking it now, you can immediately satisfy any desire that $100 can cover.

This concept can be applied to the privacy paradox. Let's imagine a thirteen-year-old boy who has just signed up for Instagram. He uploads hundreds of contacts, including school friends from his phone, and discovers that most of them have Instagram accounts. He follows them all, one by one. Dozens of accounts start populating his feed, and within minutes, he's checking out the latest photos shared by his friends, his crushes, and kids from the neighboring town. The satisfaction of these social interactions—the immediate gratification offered by every one of those frictionless taps and swipes—is palpable. He's young, after all, still a middle schooler. He is suddenly engaged with a world of people he might otherwise never have known. The possibilities feel—and truly are—endless.

When he signed up to Instagram, though, he did not consider the implications to his privacy—in part owing to the framing of the decision put before him. He was probably unaware that Instagram can discover who his friends and family are, the types of media music he enjoys, and whether he is a bully, or has been bullied, and can predict his future political leanings and even what kinds of girls he likes. Nor did he consider the possibility that Facebook's servers might be breached, or even worse, that some future Cambridge Analytica–connected foreign academic could run an anonymous experiment on him, draw all of his and his friends' information out of the platform, and sell it on the open market to unscrupulous governments and corporations.

This example illustrates the privacy paradox: there are huge positive returns from joining the service now, but there is little thought given to issues of privacy; we engage in what the MIT neuroeconomist Drazen Prelec calls a "hedonic relationship."[19] This is the irrationality of maintaining a "present bias," as dictated by hyperbolic discounting, at work. Unaware of the inner workings, the user has not considered the potential negative payoffs. Even if these concerns are understood, the overall magnitude of those future payoffs is uncertain. They apparently would occur far into the future—and, accordingly, they are naturally brushed off as part of the trade-off.[20]

This hyperbolic discounting affects not only teenagers but also, as the noted behavioral economist Alessandro Acquisti has found, both the "naïve" and the "sophisticated," who "perceive the risks from not protecting their privacy as significant."[21] In other words, it affects just about everyone.

Industrial Obscurantism and the Mass Hoodwinking of Users

The privacy paradox has major implications for our country. The consistent breach of privacy, security, and the public's trust have not only forced us to lose faith in the internet industry, they seem also to have corroded the political fabric of our country itself: the norm now is for disinformation operators to infiltrate our systems. Foreign operators engage in disinformation operations because the infrastructure to spread it is readily available and there is ample economic and political incentive, and little disincentive, to do so. We should continue to study the reasons for the privacy paradox and determine mechanisms to better inform consumers and strengthen the campaign for policy changes.

Engagement in such an undertaking, however, is what I describe as systemic "industrial obscurantism."

Some scholars have been accused (by other scholars) of practicing intellectual obscurantism, using unnecessarily complex language and forms of argument to deliberately restrict the spread of knowledge. The practice is unnerving, undemocratic, and unwelcome. No one appreciates an arrogant savant—especially not one who deliberately withholds information that may be important to the public.

A related practice that I call "industrial obscurantism" occurs when industry executives use language that unnecessarily complicates interactions with users to obscure their businesses' commercial strategies to entrap an unsuspecting public.

Regrettably, industrial obscurantism has infected nearly every chief executive in Silicon Valley—and, again, Mark Zuckerberg is Exhibit A. The instances are too many to count: Zuckerberg's wish to maximize "meaningful social interactions," his company's intent to disseminate only ads that are "good" and "desirable" to the end user, Facebook's urge to "connect the world." Facebook consistently claims to represent apparent desires for positive social outcomes. But if Facebook truly wanted to ensure meaningful social interactions over its platforms, it would severely limit the capacity for hate speech and disinformation to spread by disabling its practices of behavior tracking, shuffling people into echo chambers, and curating the content it thinks we want to see in efforts to boost its digital ad space. If Facebook truly wished to allow only ads that a user finds acceptable, then it would do a better job of asking users about what kinds of ads they want to see and reflect those preferences in the users' news feed. And if Facebook truly cared about connecting the world, the company would not pursue build-outs of digital and physical infrastructure that serve only the company's profit-maximizing objectives—as in the case of its iterations of zero-rating programs, including Facebook Zero, Internet.org, and Free Basics, through which the company seeks commercial ties with telecommunications providers (mostly in the developing world where few can afford internet access) and subsidizes the use of a set of services that primarily feature Facebook's own platforms at no cost to the user. Chile's Subsecretaría de Telecomunicaciones and the Telecom Regulatory Authority of India, which possess the governing means and political will to reject such proposals, have asserted that these programs violate net neutrality conditions and have accordingly banned them. The situation they wish to avoid is clear: as Indonesian researcher Helani Galpaya found, "11% of Indonesians who used Facebook also said they did not use the internet." Facebook knows very well that pushing zero-rating programs on the developing world is in the long-term commercial interest of the company; as soon as the network effect is established and the services' adoption curve hits its point of inflection, Facebook itself will become what is regarded as the internet by the local population. Indeed, it has been found that the majority of users in Nigeria (65 percent), Indonesia (61 percent), India (58 percent), and Brazil (55 percent) believe Facebook *is* the internet (in contrast with 5 percent of U.S. respondents).[22] And while Zuckerberg does not confess to the highly commercial nature of Internet.org's motivations, suggesting only that "if you do good things for people in the world, that comes back and you benefit from it over time,"[23] his chief financial officer—who is perhaps more compelled than others to concede any ties to profit-seeking given his role—has noted incisively that "focusing on helping connect everyone will be a good business opportunity for us, [and] as these countries get more connected, the economies grow [and] the ad markets grow."[24]

Perhaps the industry's most notable use of industrial obscurantism is Zuckerberg's announcement in early 2019 that Facebook would be making a "pivot to privacy," which caused apparent confusion among the journalistic community and consternation on Wall Street.[25] The proclamation's appeal to the general global public is apparent. But where in actuality is the commitment to privacy when considering the tracking-and-targeting business model that is the cornerstone of its commercial success? Why indeed would Zuckerberg subvert the company's entire value proposition, which has made his shareholders a mint, unless he were compelled to do so by the government? Predictably, commentary and analysis of the proposed "pivot" fell on both sides: some suggested that he truly cared about privacy and that this would represent a sea change for the company, while others were anything but hopeful.[26]

The latter camp was and remains correct. What Zuckerberg means to do in making this pivot to privacy does not at all entail ceasing Facebook's tracking-and-targeting operation—a business practice that deals in tremendously high margins and is not subject to any form of accountability or regulation. The greatest privacy concern for Facebook users—the concern that users might even have thought Zuckerberg was referencing in discussing the pivot—lies in this operation. But it is this operation that almost unilaterally constitutes the company's value proposition as well. When you possess a vertically integrated monopoly like the Audience Network, a platform that you designed over the past decade and that generates fortunes for your shareholders, you do not yield it by voluntarily committing to privacy protections to the public that would slash at your company's knees—not even if the president of the United States himself were to unilaterally proclaim he was going to come after you. In this conceptualization, is it not the case that Facebook is working to ensure that "the future is private"[27] from third parties like the government rather than from Facebook?

Instead of shifting existing privacy-invading practices toward a more sensitive and sensible approach, all the company will actually do, at least in the term over which the proclamation applies, is integrate additional new services that some might say are more privacy sensitive, such as Facebook's new model for engagement through its Groups, or its ongoing shift toward end-to-end encryption throughout the company's services.

This is raw, industrial obscurantism. The public explicitly cares about only one kind of privacy in its dealings with Facebook: privacy that enables consumer choice in a way that can protect us from such negative externalities as disinformation campaigns and foreign interference in U.S. elections. This desire can only be satisfied by dissecting the company's business model, breaking it down into its constituent parts, and reassessing them from head to toe. Zuckerberg is knowingly addressing a different sort of privacy—protection of our communications from unwanted hackers and potential third parties. We of course care about that kind of privacy too; however, that is not the kind of privacy we are calling for in the Facebook of today. This is a subtle but effective Alice in Wonderland approach to the meaning of "privacy" in the extreme. That said, we will not be receiving any additional privacy from the peering eyes of Facebook itself unless it is compelled by the federal government or intense public pressure. Facebook's high-walled garden will continue to protect the company's valuable vertical integration of platform media and ad targeting for the foreseeable future.

One could also examine this from another angle: if Zuckerberg truly were talking about breaking down the business model in any way, then it would actually be in his interest to disclose that voluntarily because doing so would obviate all of the inevitable bad press he received for prioritizing the misleading pivot. But he did not. Even as a scrutinizing society, we walk away from this exchange thinking that the executive cares about user privacy and will go to great lengths to protect it. Nothing could be more untrue.

But again, these suggestions of course depend upon how one defines privacy in the context of Facebook.

Zuckerberg is not alone in engaging in industrial obscurantism. The same arguments can be made about Google's commitment to consumer privacy as voiced by Pichai. Pichai might say all of the right things—in particular, that he wishes for Google to develop strong privacy choice architectures for consumers throughout Google's products, as he suggested in a recent *New York Times* op-ed.[28] One can be certain, though, that the core data-collection and data-processing engines that power Google's commercial infrastructure will remain intact absent an immediate regulatory threat from the government—or more precisely, until such time that it no longer serves the company's long-term commercial interests to pursue them. The industry publicly suggests that it hopes for a strong U.S. legislative approach to tackle privacy, while privately suggesting to policymakers that any federal law needs to preempt the states' power to innovate any further through novel regulation. Such national preemption works in the industry's favor because it has become remarkably difficult to adopt any changes into law. And that incentive for preemption is a precondition that internet trade associations will advocate for in closed-door meetings in the halls of Congress.

Tim Cook, too, has engaged in rank industrial obscurantism. In recent years he has become famous for pointing the finger at companies such as Facebook and Google, arguing that the development of a business model engineered on the strength of knowing anything and everything about each and every individual consumer is corrosive to society. He claims that Apple collects far less data, and more critically, that Apple's business model is not premised on the forms of customer exploitation exhibited by the leading internet companies. Perhaps Apple has not exploited the American consumer to the extent that Google, Facebook, Twitter, and Snapchat exploit the data associated with their users. But, as noted in chapter 1, Apple has no qualms about trampling over the human rights prospects of Chinese citizens. An intellectually honest executive cannot in one moment finger Facebook for engaging privacy harm in the United States and in the next turn a blind eye to his own company's facilitation of violating the human rights of an entire country.

Apple could argue, for one, that it would not provide back-door access to the Chinese government. We have little to go on to verify that fact, other than the company's word. And in any case, we cannot be sure that the company can technologically live up to this commitment in the face of a coercive government that is said to employ fleets of brilliant hackers. Once Apple has gone down the path of quiet compliance, the power in its relationship with the Chinese government sits completely with the Chinese. Many leading journalists have conducted interviews with Tim Cook in recent years. I would like at least one of them to ask the following: You argue that Facebook's data practices implicate the public interest by perpetuating such harms as disinformation, and that we need better protection from Facebook's business model.[29] But does Apple not also put business over ethics when push comes to shove—such as in the case of your compliance with the Chinese security regulations? Did not Apple comply in its own business interests and at the expense of Chinese citizens' personal privacy and human rights?

It is in Apple's commercial interest to criticize Facebook; Facebook is cash-rich and invests in technologies that could one day threaten Apple's hegemony over the closed-OS mobile-phone market. This is why Cook pushes for privacy regulation for Facebook—to eat into Facebook's record profit margins and force Facebook to be more conservative in its technology investments, which will otherwise increasingly challenge Apple's position in the technology markets.

Why is it so difficult for the leading Silicon Valley executives to escape engaging in industrial obscurantism? Because it is in their nature to put lipstick on a pig. Today's technology executives are seldom different from bankers or oil profiteers: they must seek profit maximization in the interests of their investors, including, in the case of public companies, their stockholders—by law. We cannot expect anything else; it would be unnatural—and, in fact, illegal—for the technology executive to place the interests of the public over those of the company. It could be argued that there is nothing wrong with this kind of behavior. After all, this is simply an exhibition of the American economic design at work. We have a free market economy; if that means that our leading firms will mobilize lobbyists and policy staff and corporate executives alike to influence Congress and the American people, then so be it. Everyone should have a voice, even companies that have only dollar bill signs in their sights. Unless we break from our current mode and adopt a new approach to our governance of for-profit institutions, as was recently suggested by chief executives of major U.S. companies who are members of the nonprofit Business Roundtable, the predilection toward industrial obscurantism shall persist.[30]

So we cannot expect anything else of Tim Cook. It must be in his nature to optimize Apple's interests at any cost. And in his case, the decision to quietly comply with the Chinese law is obvious. He must, to protect the market interests of his company and thereby his shareholders; failing to comply with the Chinese law or voluntarily divesting Apple's Chinese interests would deal a substantial blow to the company. Our goal must be to drain this industrial swamp of exploitation by seeking a new economic circumstance for citizens through fair regulation. But we cannot; because of obscurantism, we the general public are up to our necks in alligators.

All of this converges on one conclusion: when it comes to consumer interests such as privacy and individual autonomy, the time has come for us to discount the perspectives of the executives of Silicon Valley. Recent history illustrates as a general matter that they speak in the interests of their companies, not those of the public.

It is time to listen to those who consider consumers' economic situations first and foremost.

The Terms of Disservice

One of the most invasive practices of companies in collecting personal information over the internet is to require the user to accept the service operator's terms of service regarding data collection, privacy, and advertising.[31] As certain platforms have become increasingly vital to participation in employment, education, and social life, these take-it-or-leave-it contracts—which have instigated new legal complaints against Google and Facebook, with European citizens enforcing their rights under the General Data Protection Regulation[32]—become especially damaging to both the consumer and the broader public interest. This is particularly true given the anticompetitive nature of the industry and its consistent policy failings—from the spread of disinformation to the constant cadence of security breaches across the sector.

Digital companies choose this design for commercial reasons. They know well that their users are likely to accept the terms of service so they can join their friends on the network, access the app's ride-hailing service, and send and receive e-mail with colleagues—and there is no better option available in the marketplace. For companies, the data collection is nominally necessary to operate their services. More important to them is the capacity to generate sophisticated behavioral profiles on the user—profiles designed for use in the digital advertising economy.

Consumers are typically unaware of the commercial regime behind the terms-of-service contracts presented to them, but they actively participate in it in two key ways: first, by generating ad space for the service provider to aggregate and sell to the highest bidder; and second, by allowing the service provider to collect personal information on an uninhibited basis. The economic logic between the individual consumer and the service operator is price inelastic; as soon as consumers click "Accept" and begin using an app, they have completed the necessary steps to use the service. Meanwhile, the service operator can charge the consumer an arbitrarily high "price" behind the veil of the service's front end, in the form of a nontraditional currency metric: a complex combination of the consumer's attention and personal data, which is collected at extortionate rates on the consumer-facing side of two-sided platforms such as Amazon and translated into dollars on the other side as revenues are collected directly from digital advertisers. The consumer's relationship with Amazon, Google, and Facebook is thus exploitative because the market offers no alternative to the necessary condition of offering the firms full access to the consumer's digital personhood—including personal information—in exchange for access to the service. In the absence of real competition, these firms do not hesitate to collect monopoly rents.

A better design for the consumer would be far simpler: offer the user the option to join the service without data collection. Instead of forcing this exploitative price-inelastic deal on users, service providers could allow consumers to opt out of data collection and advertising—or, better yet, they could offer an opt-in to data collection and advertising on a more granular basis. This would shut down the service provider's

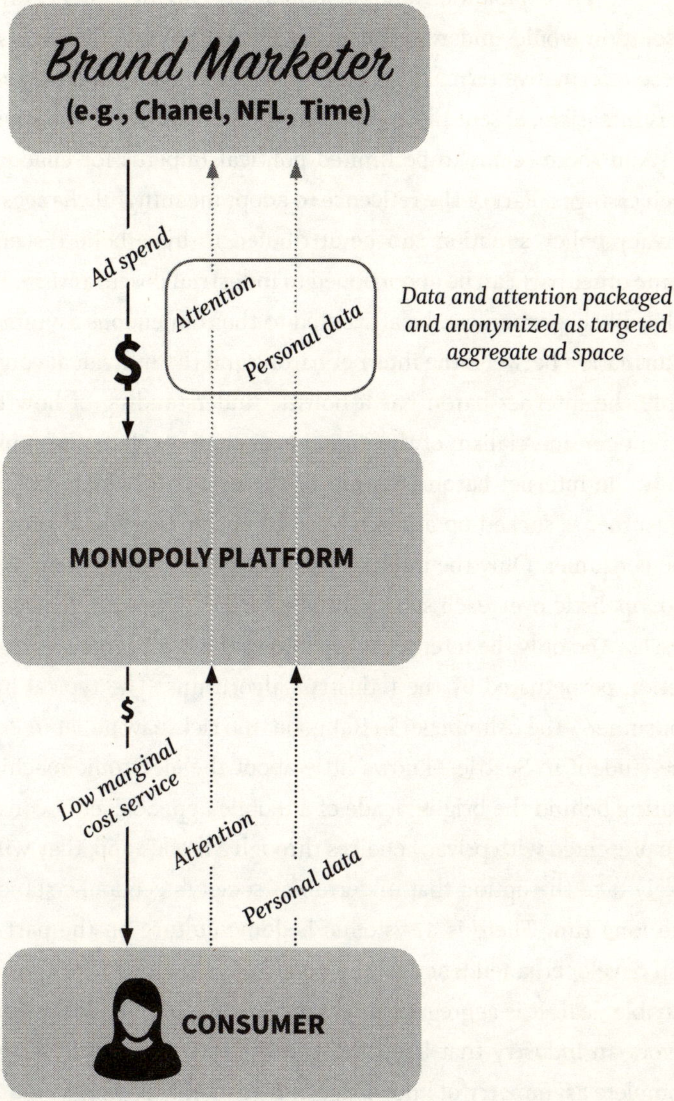

The complex combination of consumers' data and attention carries high commercial value to the monopoly platform, which anonymizes and aggregates it in service of engaging brand marketers, who wish to influence consumers' behavior.

capacity for exploitation and monopoly leveraging.[33] Crucially, such a solution would undercut the firm's veiled business model. As such, these reformative terms of service will not emerge voluntarily and are thus infeasible, absent the appropriate governmental interventions.

And there seems to be limited political impetus for enacting any such changes. Part of the reticence to adopt meaningful changes to our privacy policy situation can be attributed to hyperbolic discounting. Some other part can be apportioned to industrial obscurantism. But another likely portion can be attributed to the tremendous asymmetry of information between the internet barons and the individual consumer. Only the internet baron has a holistic understanding of how the inherent commercialism of the digital ecosystem exploits the consumer. Only the internet baron is aware of the extent to which data on the consumer is sucked up and analyzed to enable behavioral profiling of the consumer. Only the internet baron is aware of the extent to which monopolistic overreach subverts the interests of open democratic processes. And only the internet baron knows the level of incisive discrimination perpetuated by the industry's algorithms. The typical internet consumer—the fishmonger in Rangoon, the rickshaw puller in Bombay, the student in Seattle—knows little about the electronic machine operating behind the bright façade of a mobile app. So even if consumers are presented with privacy choices through a mobile app, they will most likely take the option that undermines society's economic standing in the long run. There is a systemic hedonic culture on the part of the app developer, a tendency to knowingly offer the unaware consumer a terrible deal. It is aggregate hoodwinking on a grand, global scale that favors an industry that is on the winning side of a battle featuring a complete asymmetry of knowledge and economic power.

These terms of disservice do not appear to be changing anytime soon.

The Streams of Data Fueling Internet Commerce

As we consider the structural overreaches of the consumer internet regarding user privacy, it is useful to understand the variety and richness of the types of data that companies collect on us to illustrate the industry's long reach into our private lives. Two types of collected data are especially sensitive and warrant particular protection. The first, personally identifiable information (PII)—the set of descriptive facts that point uniquely at a person's identity—is especially sensitive and as such has traditionally received heightened protection from governments. The second type of sensitive data—biometric identifiers—are less well known. Much like PII, biometric data can be used to verify that you are who you say you are and, similar to PII, can link your identity with your activity across the web and in the real world.

Personally Identifiable Information

PII typically includes such unique identifiers as a person's name, e-mail address, phone number, Social Security number, tax identification number, employee identification number, and other unique signifiers. Given the sensitivity of a unique identity, personal identifiers have had separate and more stringent protection in most U.S. laws that address consumer privacy and data security. The California Notice of Security Breach Act, for example, defines "personal information" as any of the following:[34]

> An individual's first name or first initial and last name in combination with any one or more of the following data elements, when either the name or the data elements are not encrypted:
>
> (A) Social security number
>
> (B) Driver's license number or California identification card

(C) Account number or credit or debit card number, in combination with any required security code, access code, or password that would permit access to an individual's financial account
(D) Medical information
(E) Health insurance information
(F) Information or data collected through the use or operation of an automated license plate recognition system, as defined in Section 1798.90.5

Legislators have determined over the years that personal identifiers deserve extra protection to prevent identity theft. Access to your name, e-mail address and password, and phone number could be used to impersonate you, including making transactions using your credit card, voting in your place, or committing fraud or other crimes under your identity. Consumer internet companies, which typically want to collect your PII so that they can connect your profile with what you do across the internet and in the real world, accordingly abide by those laws and take steps to protect the security of your personal identifiers against breach.

It is not necessarily a bad thing that consumer internet companies collect our personal identifiers. Internet companies are incentivized to ensure protection of personal data to the utmost and employ a fleet of talented security engineers to that end. They collect information primarily to verify our identity and to link our accounts with our other experiences across the web—a reasonable practice, one could contend.

But the legislative and legal traditions of treating PII as more sensitive than other forms of customer data (such as the behavioral data Facebook and Google collect) have categorized this information into different realms of cybersecurity and privacy practice. The requirements for protection of non-PII are weaker than those for personally identifiable information; the latter enjoys more stringent and prescriptive protection under most laws. An industry lobbying culture ensures that both state and federal legislatures perpetuate this dichotomy, preventing any real accountability on all forms of data that are not PII. While they keep legislatures focused on regulating PII, companies such as Google can continue undeterred as they collect other, more sensitive forms of personal data on consumers' personalities and behaviors. Meanwhile, they can maintain an appearance of responsibility, since they are advocating for a privacy law of some kind in pushing for PII regulation. Unfortunately, the fact that these laws have almost no business impact on companies like Google is all but forgotten—or goes ignored—by legislatures.

This tendency has been a failure of our legislatures. It has enabled the Cambridge Analytica incident, which was focused on the collection of behavioral and usage data on Facebook users, as well as the psychological inferences built on top of that raw data set. And while some legislatures are now pushing for definitions of PII to be updated to reflect the burgeoning collection of information on consumers in every day digital life—California, to name one, requires protection of any data that is linked to the consumer's identity—it will remain a noteworthy political challenge to push for progressive changes in how non-PII information is protected.

We need to treat behavioral data collected on internet users as just as important as our PII. Until our laws and legal regimes reflect this reality, we cannot hold the internet companies accountable; to date we have not even targeted the non-PII data that powers their implicit business models.

Biometric Identifiers

Much like PII, biometric data can be used to verify that you are who you say you are and, similar to PII, can link your identity with your activity across the web and in the real world. But there is a critical distinction between biometric identifiers and personal identifiers: biometric identifiers are a creation of technology and your biological makeup, whereas PII is a human creation. The latter is subject to change, at the will of the data subject. You have autonomy over your PII because your contact information can be readily changed, and your other identifiers live simply as numbers and letters on a document. Meanwhile, biometric identifiers are typically constant for life. You cannot change the look of your iris, your fingerprints, or your DNA. And you cannot change a consumer internet company's facial-recognition metric designed to model the unique physical structure of your face.

The distinctive quality of biometric identifiers carries massive implications for the individual consumer but remains little understood. Biometric identifiers—and facial-recognition metrics, in particular—enable powerful internet companies to engage in seamless tracking of your activity across the web and the real world. Internet companies need not cross-check your name or e-mail address with other service providers; they can simply implement off-the-shelf facial-recognition algorithms on an image they might find somewhere on their platform, on a third-party website, or as a result of their partnerships with brick-and-mortar stores that take pictures and videos of customers who enter their establishments. Subsequently, they can infer that you were responsible for that activity, through the photo itself and through triangulation of other aspects of the photo—including who shared it, where it was taken, when it was taken, who else was in it, and so on. This represents a startling invasion of privacy and an industrial culture of radical anonymous tracking that consumers are practically unaware of. Collectively, we are left in the dark.

The creepiness of being tracked by a consumer internet company that uses your facial-recognition metric has reached far into the physical world. Already, retailers and marketers are developing artificial-intelligence systems that can be used to infer what shoppers are thinking and feeling as soon as they enter a store or pick up a particular product off the shelf.[35] In fact, the American Civil Liberties Union found in early 2018 that Lowe's Home Improvement was already using facial recognition, without user consent, to identify shoplifters, while most of the other retailers they queried refused to answer.[36]

The industry will inevitably link facial-recognition metrics drawn up in brick-and-mortar stores with parallel metrics developed online through analysis of photos shared on sites such as Facebook and Google—as well as those traced from across the web. The algorithms implemented in physical shops could someday even be licensed out to store chains by the internet companies themselves, in relationships similar to those that Google has established with Starbucks to offer free Wi-Fi. [37] In the future they might choose to offer free access to the software without requiring licensing.

Internet companies are developing facial-recognition technology and implementing it over their services for a second but no less important reason: to maximize your engagement with their platforms. Think of your use of Facebook or Instagram. When you see untagged friends in your photos, the service recommends you tag them. It might even suggest which of your friends it is. That reduces a critical transaction cost: you do not have to type in the person's name. That bit of commercial edge in engagement is precisely what the internet companies are on a continual quest to optimize.

Nearly every major consumer internet company is developing its own proprietary facial-recognition software. Facebook, Amazon, Microsoft, Google—all of these have developed and continue to refine highly sophisticated and remarkably accurate facial-recognition technologies that can be implemented for photos, videos, and other novel media. A notable indication of the consumer internet companies' keen commercial interest in maintaining unregulated use of facial-recognition technology is their vociferous opposition of any sort of law that would curb their potential to implement the technology. The technology sector's heavy lobbying in opposition to the Illinois Biometric Information Privacy Act is illustrative of this.[38]

Web Browsing and Mobile-Use Data

Imagine anonymous companies having a complete view into everything you do over the internet—which websites you spend the most time on; which sports articles, investment brochures, fashion lines, or architectural designs you find the most interesting; which news and media outlets you read the most. All of the top-level domains you visit—and their political leanings—as well as the behavioral inferences associated with your on-site interactions and historical browsing activities are available to internet companies.

This is the reality in which we live today. How can this be? The answer lies in the development and ongoing refinement of a highly effective set of web technologies that the leading internet firms have systematically and silently installed throughout the open web to infiltrate your life and spy on you to maximize value out of their digital ad space and targeting capacities.

While the web technologies that enable this level of surveillance on consumers are many, the most critical to the backbone of the consumer internet is the seemingly benign web cookie.[39] Cookies are snippets of code injected into webpages that track your activity on those webpages—all to inform either the webpage owner, an internet company, or some anonymous third party.[40] They are seemingly everywhere. One recent report published by the National Endowment for Democracy's Center for International Media Assistance found that of the fifty independent publishers researched, 92 percent featured third-party trackers, including cookies and web beacons, which are used to monitor the activity of web users for the purpose of web analytics, page tagging, or email tracking.[41]

Why would these various types of firms want to know about your behavior on the page? It might be that the webpage owner wants to maintain knowledge of your past behaviors so that it can tailor the site to your personal preferences. For instance, a news website might want to maintain easy access to your location so it can send you local stories, or a search engine might want to understand your ethnicity and cultural preferences to tailor your search results, or an e-commerce platform might want to be able to maintain your online shopping cart.[42]

But the more objectionable use of web cookies is our real concern here. This use involves web tracking to engage in the collection of data on the individual, all toward the end of tracking his or her personality for commercial purposes. For instance, a news site like *The Epoch Times* might track your activity on its site to infer your political leanings and maintain a list of different classes of conservatives who align with specific, narrower conservative camps. Those lists would then become immediately useful and therefore highly valuable to *The Epoch Times* in its marketing activities, as well as to other conservative media outlets, political candidates, political parties, and firms that sell certain kinds of products and services. This might indeed be precisely the way that Epoch channeled the pro-Trump advertising it pushed on Facebook via shell groups carrying harmless-sounding page names such as "Honest Paper" and "Pure American Journalism," which prompted a takedown by Facebook in August 2019 but only well after the fact and much too late.[43]

Why would website operators collect this kind of information in the first place—especially if they do not directly sell your information to other firms? (This is what, for instance, many newspapers tend to promise—that they absolutely will not sell or share any information they collect on you to third parties.) The reason is actually quite natural: to maintain their current reach and increase their potential audience. But this urge to maintain an audience and stay competitive in an evolving media environment where every outlet has to vie for the discerning customer's attention has forced the hands of news sites like the *New York Times* to subtly subscribe to and empower the ongoing hegemony of Facebook and Google.[44] Even such benign intellectual outlets as Scientific American, which notes that it implements cookies "to personalize the website for you, including targeting advertisements which may be of particular interest to you," are capable of engaging in such practices.[45]

The finest illustration of this is the Facebook pixel.[46] Like standard web cookies, the pixel is a snippet of code that Facebook encourages websites to install. It is all made very simple—the code is already developed, and as a website operator, all you have to do is copy it into your site code. The pixel—an actual one-byte-square pixel that contains a vicious set of lines of code within—then unleashes itself on the website as soon as the consumer loads the page. Instantly, through this surveillance tool, Facebook has gained access to your browsing information on whatever third-party website you are visiting. But most critically, the installation and use of the pixel also enables the website operator to more effectively track and target its users on Facebook's media and advertising platforms—and that is what is so valuable to the operators of leading media companies such as online newspapers.[47] The company needs to be able to reach its consumers over the platforms on which they spend the most time, in an intelligently targeted manner. In this eternal competition over customer attention among media properties new and old, every additional bit of information they possess on established and potential future subscribers can be helpful to commercial sustenance and corporate growth.

The irony is that to compete with other outlets that vie for consumer attention (for example, website loads and subscriptions), even organizations such as Consumer Reports, a consumer advocacy organization that vociferously fights for online consumer privacy protections, are silently but surely coerced into installing the Facebook pixel on their websites.[48] A bidirectional exchange takes place—one that leverages an extraordinarily powerful but hard-to-see network effect. The website operator gains the ability to target ads at the people who have visited its website over Facebook and over Audience Network, Facebook's tool that enables use of the company's advertising platform to place targeted ads on third-party websites—sites that operate independent of Facebook.[49] The website operator can inject data gathered through both first-party cookies (which are owned and operated by the website itself) and third-party cookies (which are injected into the website by other companies that are neither the website operator nor Facebook but some other entity the website operator has arranged an information exchange with for commercial purposes) installed on the site and send all of that to Facebook, via the pixel, to engage in more sophisticated targeting online. Facebook specifically encourages the use of first-party cookie data in this injection, as it notes in its brochures touting the functionalities of the pixel to digital marketers:[50]

> You can now use both first- and third-party cookies with your Facebook pixel. The difference between first- and third-party cookies lies in who owns the cookie. First-party cookies are owned by the website a person is currently viewing, while third-party cookies belong to a website other than the one a person is currently viewing. Compared to third-party cookies, first-party cookies are more widely accepted by browsers and stored for longer periods of time. To give you more control over your advertising outcomes, the options for using cookies with your Facebook pixel are as follows:
>
> (1) *Use the Facebook pixel with both first-party and third-party cookies.* This is the default option and is most likely your current Facebook pixel setting. With this option, you will use first-party cookie data with your Facebook pixel, in addition to third-party cookie data. Using both first-party and third-party cookies will enable you to reach more customers on Facebook and to be more accurate in measurement and reporting.
>
> (2) *Use the Facebook pixel with third-party cookies only.* You can disable first-party cookies and use the Facebook pixel with third-party cookies only. With this option, your Facebook pixel will be less effective in reaching customers on Facebook and less accurate in measurement and reporting.

Facebook encourages website operators to adopt the Facebook pixel to expand and consolidate its power across the open web as part of a broad effort to drive marketers onto its platform and collect vast quantities of data on customers even while they are visiting a non-Facebook website. The pixel is Facebook's *sine qua non*—the single data-raking web technology that at once buttresses the company's unique value proposition by forcing advertising clients to use Facebook's proprietary ad management systems and maintains the company's vast visibility into the universe of commercial interaction over the web—and by corollary, the trajectory of all commercial activity itself.

Websites use the pixel just to stay relevant to their current and potential customers. This is why even the media outlets that were most critical of the company in the aftermath of the Cambridge Analytica revelations—including the *New York Times*, the *Guardian*, the *Observer*, *WIRED*, and *Recode*—have all implemented the pixel to understand their users and target them over Facebook's platforms. The economic incentive for marketers to engage Facebook is overpowering: a marketer that wants to stay in business in 2020 has no choice.

Google has a series of analogous tools and technologies to maintain a mind-numbingly invasive level of reach rivaling that of Facebook. Some progressive advocates have attempted, largely unsuccessfully, to defeat the hegemony over the tracking-and-targeting economy dominated by Facebook and Google and reflect greater transparency to the end user. Perhaps the most notable is legislation that would operationalize a "Do Not Track" regime—a set of actions by which consumers could dictate through their web browser that they wish not to be tracked by the website in question, a preference the website would have to respect.[51] Despite the failure of such legislation to pass Congress, some of these efforts have nevertheless resulted in web technologies that consumers can download, install, and use to block participating firms from acquiring browsing data on them. Google, for instance, enables users of the Chrome browser to send a do-not-track request "with your browsing traffic."[52] Tools that help users block some digital ads, web cookies, and trackers, like eyeo's Adblock Plus[53] and Electronic Frontier Foundation's Privacy Badger,[54] also exist. These technologies, however, remain either unknown or underused by the vast majority of internet users, and even when they are installed, they are only effective at blocking the peering eyes and ubiquitous reach of the largest internet companies. The news that Google will work toward shutting down third-party cookies over its web browser Chrome, which leads the market, and limit cross-device tracking through enforcement of its new SameSite rules is also not necessarily encouraging. This corporate policy—which is best described as a business strategy—could well serve only to lock down Google's domination over behavioral profiling. This in turn may hurt market vibrancy in the digital ad ecosystem, implying that Google could grow into an ever-stronger position over time.[55]

On-Platform Engagement Data

Every time you open the Twitter app you are delivering engagement data straight to Twitter's servers. Our activity on our Twitter feed, where we hover on certain posts, liking and retweeting a number of them; our click-throughs to the articles and third-party websites that we like, where we might spend several minutes exploring; the tweets we put out ourselves, including the articles that we liked or the tweets that we found inspiring; and the fact that we opened up the Twitter app in the first place—these are all examples of engagement data, information that the company uses to understand who we are and what drives us.

Engagement data can easily be used to infer behavioral characteristics about us. This kind of information is rich; companies with the cash flow possessed by the likes of Twitter have each developed tremendously powerful algorithmic machines to keep us engaged on the platform. There are clear benefits to marketers who have access to social media companies' knowledge of our behavioral characteristics. Professors Dokyun Lee, Kartik Hosanagar, and Harikesh Nair have conducted research indicating that well-designed marketing content can generate "immediate leads (via improved click-throughs) with brand personality-related content that helps in maintaining future reach and branding on the social media site (via improved engagement)."[56]

This engagement data is limited not only to internet companies. For instance, a robust industry of data harvesting has sprung from producers of smart televisions and their media partners, as researchers have recently documented.[57] Professor Arvind Narayanan of Princeton University notes that, "When we watch TV, our TVs watch us back and track our habits. This practice has exploded recently since it hasn't faced much public scrutiny."[58] The insights drawn from such data can readily be injected into data markets and used to manipulate the organic content and ads you see on the internet.

But this level of surveillance is dangerous. It involves the collection of engagement data, which, in the case of YouTube, is expressed through your viewership of and interaction with videos and enables YouTube to determine your underlying psychology—without your understanding and awareness. The only goal of the platform and the sole purpose for its existence is engagement maximization—to get people onto the platform and to stay there for as long as possible to maximize ad space that can be sold off to the highest bidder. This underlying goal is also at the root of a serious issue whereby users have been pushed by the YouTube platform to view content that is hateful, conspiratorial, or just plain false and misleading.[59] (Chapter 3 more fully discusses the issue of engagement maximization.)

Social-Graph Data

We all communicate with some number of friends, acquaintances, family, colleagues, and others. Our social graph is the network representation of all of those individuals. In the social media context, this might include all of our friends as well as any connections our friends might have. The case is similar for telephone contact lists, e-mail contact books, and so on.

Every person has a different social graph. We each communicate with a different set of people in the physical and digital worlds. Those communications are represented through our true social graph. On first blush this might not seem to be that valuable: what can information about your network of friends really tell internet companies?

In fact, quite a lot. Through our social graphs, internet companies can infer our behaviors, preferences, interests, beliefs, and other characteristics about our true selves. We tend to be homophiles: we associate with things and people that are like us. Miller McPherson, Lynn Smith-Lovin, and James Cook put it well: "Similarity breeds connection. This principle—the homophily principle—structures network ties of every type, including marriage, friendship, work, advice, support, information transfer, exchange, comembership, and other types of relationship. The result is that people's personal networks are homogeneous with regard to many sociodemographic, behavioral, and intrapersonal characteristics."[60]

There are inevitable exceptions to this rule, but on the whole students at a university are closest in the real world to other students at that university and employees who communicate with one another at a company are usually closely affiliated. The more representative a social graph the internet companies can obtain on us, the more behavioral characteristics about us they can infer. It is yet another powerful capacity that consumer internet companies possess to cross reference and confirm our behavioral traits by checking those of our friends. Researchers have proven this, including Nathan Eagle, Alex Pentland, and David Lazer, who found that 95 percent of friendships can be accurately inferred using observational network data from mobile phones.[61]

Another aspect of social-graph data that is important to consumer internet companies is engagement maximization. In the same way that we associate with those we are like, we find social media posts from people we like to be especially engaging. The more the companies can leverage our social-graph data to feed us engaging material, the longer we spend on their platform. The raw commercial power that comes with the knowledge of our social-graph data is so desirable that companies operating web-based services that include mobile applications often push us to disclose our smartphone contacts to them—often, they claim, to enable technical functionality of the service. While that may be true, the commercial purpose for that data collection is nevertheless real. Facebook researchers have noted that "newcomers who see their friends contributing go on to share more content themselves. For newcomers who are initially inclined to contribute, receiving feedback and having a wide audience are also predictors of increased sharing."[62]

Companies that are able to obtain access to your social-graph data—like Zoom, which in the course of the coronavirus outbreak was vilified for implementing software that was sending its users' data to Facebook—can circumvent the need to find out about you through other means. You might not be a user of Facebook's services, but if they are able to gain access to your contact list—and particularly the people that you communicate with the most—they have obtained enough to effectively target you with personalized advertisements and content over third-party platforms. Many academic results have been published on the intellectual interests shared by close contacts. Xiao Han and her colleagues verified the "homophily of interest similarity across three interest domains (movies, music and TV shows), . . . [revealing] that people tend to exhibit more similar tastes if they have similar demographic information (e.g., age, location), or if they are friends."[63] Overlaying this result on WhatsApp offers a good example: Facebook might possess relatively little information about you generated over its main service if you rarely use it, but through analysis of your social graph, which you reveal through your use of WhatsApp (including by sharing your personal contacts), Facebook can technically infer your likes and dislikes based on your closest friends' preferences and thereafter target you with behavioral advertising through its Audience Network, which places targeted ads at Facebook's determination on third-party websites that you might frequent. While the public lacks transparency into Facebook's current practices concerning data-sharing arrangements with WhatsApp, reports in 2016 that Facebook would begin using

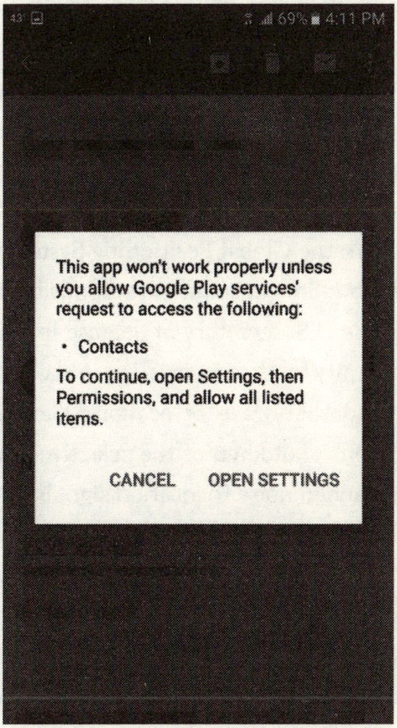

Google Play application downloading and purchasing platform requesting access to the user's phone contacts.

WhatsApp data for ad targeting—despite WhatsApp's pre-acquisition commitments—are portentous.[64]

Location Data

Users' location has such tremendous value in the digital ecosystem that nearly every modern consumer brand company tries to maintain precise knowledge of it. A number of different technologies, mostly available over your smartphone, offer the industry a detailed look into where you are and when you are there. The most obvious source of data on our precise geolocation is information obtained primarily through our communications with the Global Positioning System (GPS) on our cell phones. When the satellite constellation responsible for giving us GPS was opened up by the U.S. secretary of defense in 1993, it unleashed a tremendous opportunity for data innovation in the private sector, seemingly overnight.[65] Another wave of economic activity resulted from President Bill Clinton's shutdown of the selective-availability program that had previously added noise to location signals to curtail the extent to which enemies of the American military could benefit from GPS technologies.[66] The next several years saw the meteoric rise of several still recognizable navigation brands, including Garmin and TomTom. As smartphones came to the fore, particularly the first iPhone, mobile-device manufacturers quickly picked up on the tremendous commercial opportunity that was so close at hand. That realization revolutionized the digital-advertising ecosystem as we know it.

But this rapid state of innovation with location data quickly led to serious questions concerning consumer privacy. In 2011, the researchers Alasdair Allan and Pete Warden discovered a disturbing file buried deep within the iPhone's data-storage vaults.[67] Called "consolidated.db," the file contained historically precise geolocation data associated with the owner's travels across England. That a company such as Apple was collecting and maintaining a rich store of detailed location data might have seemed like a surprise back then. Today, the practice is business as usual in the industry.

Accordingly, a deep-pocketed industry has risen from this newfound access to our GPS location data, which uses the triangulation algorithm implemented through the GPS constellation and Clinton's discontinuation of the service-availability military restriction. Crucially, our GPS data enable the inference of our individual interests, beliefs, behaviors, routines, likes, and dislikes. As professor Stephen Wicker of Cornell University has noted, the "increasing precision of cellular location estimates is at a critical threshold; using access-point and cell-site location information, service providers are able to obtain location estimates with address-level precision."[68] Any firm with access to our historical location can begin to infer our individuality. Our location data tell the story of where we live, where we work, how we get to work, what kinds of business establishments we frequent, which friend groups we go out with for dinner and drinks, who spends the night at our apartment, and so much more.

This is exactly why, as soon as you download almost any mobile application, whether it's Pandora or Facebook or McDonald's, nearly the first objective the app has is to gain access to your location data. The websites ask you for it, claiming that they need your information to service the functionality of the application—and then they systematically use it for gains in digital advertising and maintenance of customer knowledge. If they are brand companies—McDonald's, for example—they might use it to keep track of who is interested in their products and how they should target ads at particular classes of the consumer market. They might even choose to sell access to that data to third parties, including data brokers. If they are platform firms and media properties, such as Facebook or the *New York Times*, they want to know how to attract the highest possible bids for ad space. This is precisely why Wicker rightly argues for a "coarse level of granularity for any location estimate available to service providers and the disassociation of repeated requests for location-based services to prevent construction of long-term location traces." This coarseness is needed simply to allow us to maintain ownership of our digital identity.

Notably, GPS is not the only location-tracking technology that offers companies users' location information. Perhaps the most important alternative is cell-tower data—information that mobile-network companies such as Verizon and AT&T have to collect to patch users through to their phone calls. While this type of location information is less granular than GPS data, it still offers wireless-network operators access to a user's general location, which can reveal the same kinds of behavioral characteristics, albeit with lesser granularity.[69]

So valuable is your location information to digital marketers and internet platform companies, and so readily available it is in the open market, that there is a flourishing industry of collection, analysis, sharing, and sale of either the raw data associated with your location or proprietary inferences about you and your traits.[70] This industry consists mostly of data brokers that you probably have never heard of, but they silently serve insights derived from location data to the powerful corporations that can afford to pay for them.

Transaction Data

Every transaction we make has three key components of associated data: the amount we paid, the date we paid, and the business or establishment that we paid. Over the past several years, a scaled economy has begun to leverage our transaction data for use in the digital ecosystem, including involvement from the industry's biggest names: Facebook, Google, Twitter, AOL, and the data broker, Acxiom.[71]

Many of us make our transactions on one principal credit card, and it is those consumers who have access to credit cards who are the most valuable to the consumer internet industry. When we open up a line of credit, it says something about us: we intend to buy and the American financial system has backed us up. We're good for it. In other words, consumers with credit cards typically have money and therefore are the ones who digital marketers will want to reach via the consumer internet.

On the face of it, transaction data might appear to be pretty sparse and uninformative. We might only make a few purchases every day. Furthermore, we might only make certain kinds of purchases on some of our credit cards. But what transaction data tell marketers, more than any other form of our personal information, is our true intent in the marketplace. Transactions constitute market decisions that we have made to put money, not just words or thoughts for internet searches, where our mouths are. Inferring a customer's intent is like gold to digital marketers. That we initiated a Netflix account with the release of the latest season of *Sacred Games*, that we procrastinate on buying airline tickets until the day of travel, that we just bought a particular pair of Air Jordan basketball sneakers—all of these tell a story to internet companies and digital marketers that enables them to optimally shuffle us into targeting segments. Lowe's, the American hardware retailer that confessed to using facial-recognition technology, is using transaction data directly as part of its marketing efforts. *Digiday* reports that Lowe's plans to add "transaction data to customer Pinterest profiles so it can suggest products to customers as advertising . . . [which] will further personalize suggestions for customers based on data; for example, if the retailer knows a customer repeatedly purchases living room furniture, it can offer up pins that align with these interests."[72]

Google's data-sharing arrangement with MasterCard, reported by *Bloomberg* in 2018, illustrates that intent is key in this industry and that transaction data has to become a factor in maintaining the consumer internet business model.[73] As startling as the revelation may have been, the arrangement made perfect commercial sense for both companies and constituted typical business in a largely unregulated data ecosystem. It allowed MasterCard to make use of an exhaustive data set that could readily be shared with third-party companies without any strong negative consequences, while also claiming the world's most powerful internet company as a principal data-sharing client that could help establish MasterCard's data-brokering capacities—not to mention the value of the data MasterCard maintains on customers. And Google was able to substantiate to advertisers that they could achieve an adequate bang for the buck by disseminating ads over Google's platforms.

There is a long-standing tension between powerful internet companies that operate advertising platforms and marketing brands that use those platforms. Marketers spend billions of dollars on Google, but get limited information back from the platform as to how their ad campaigns performed: which audience segments saw their ads, which user groups most engaged with the ads, and which user groups are most cost effective to be targeted in the future and at what rate. Google knows the answers to these questions in absolute terms; the company is able to analyze which particular individuals receive which particular ads, how many of those ad displays lead to a click-through to the content that sits behind the ad, and—with the MasterCard data—whether those customers actually bought the product advertised. By optimizing its profits against these computations, Google has designed a proprietary and opaque path to market monopolization in services such as the internet search. By hoovering up MasterCard's transaction data, Google was able to position itself into an incrementally secure position of knowledge, affording it surveillance power over consumers and advertisers alike.

Mobile Ecosystem Data

Even if you were to remove their access to the other forms of data discussed here, the Silicon Valley internet behemoths would nevertheless be able to accumulate masses of knowledge about you through the ecosystems they develop and operate over novel connected communication architectures—in particular, through mobile-operating systems like Google's Android service and Apple's iOS, as well as digital-advertising platforms like those operated by Google and Facebook. Over the past ten years, firms such as Google have systematically made strategic moves in the ecosystem market—which, some allege, constitutes illegal, anticompetitive behavior—to ensure the rise of their ecosystem offerings and their firm's long-term commercial success at the expense of competitors.

These mobile ecosystems enjoy expansive reach. The Android operating system, for example, was installed on 2.5 billion active consumer devices in 2019, which represents the vast majority of smartphones in use in the world today.[74] Similar to Facebook's strategic connivances in installation of its pixel across the web, Google dangles direct access to end customers over its standardized Android OS, an offer it makes to mobile-phone manufacturers and network operators in exchange for their exclusive installation of its Android mobile OS on every phone they sell.[75] Of course, Android has tremendous brand value to consumers, too—we all know and love the little green droid that gave the OS an identity—but Google's monopoly power over the open mobile-OS market is undeniable.[76] Given the number of people around the world who demand an open mobile OS because of the high price of an iPhone, the potential market value to be claimed is vast.

Imagine all of the data associated with those billions of devices being compiled, stored, and analyzed by a single service operator. That is precisely what is happening. It is an incredibly rich set of data on Android customers, including their smartphone habits and behaviors, the mobile applications they download, the mobile apps they most frequently use, and the system information consistently collected by the phone and managed by the operating system. True to form, two-thirds of the data collected through Android use is gathered without the full understanding of the consumer.[77] Those data are sent straight back into Google's servers and, given the limited regulatory scrutiny over the sector, can readily be injected into the company's digital advertising regime.

Sundar Pichai's implications that his company has enabled more choice, given its free cost to customers, thus appear wrongheaded at best.[78] Chris Smith, a commentator for the Boy George Report online news site, puts it best in his analysis, noting sardonically,

> Yes, like Google could have afforded to charge a license for Android back when the company was a nobody in the mobile market. Google was smart about three things back then. One, that mobile would

Worldwide Mobile OS Market Shares (May 2016 to July 2019)

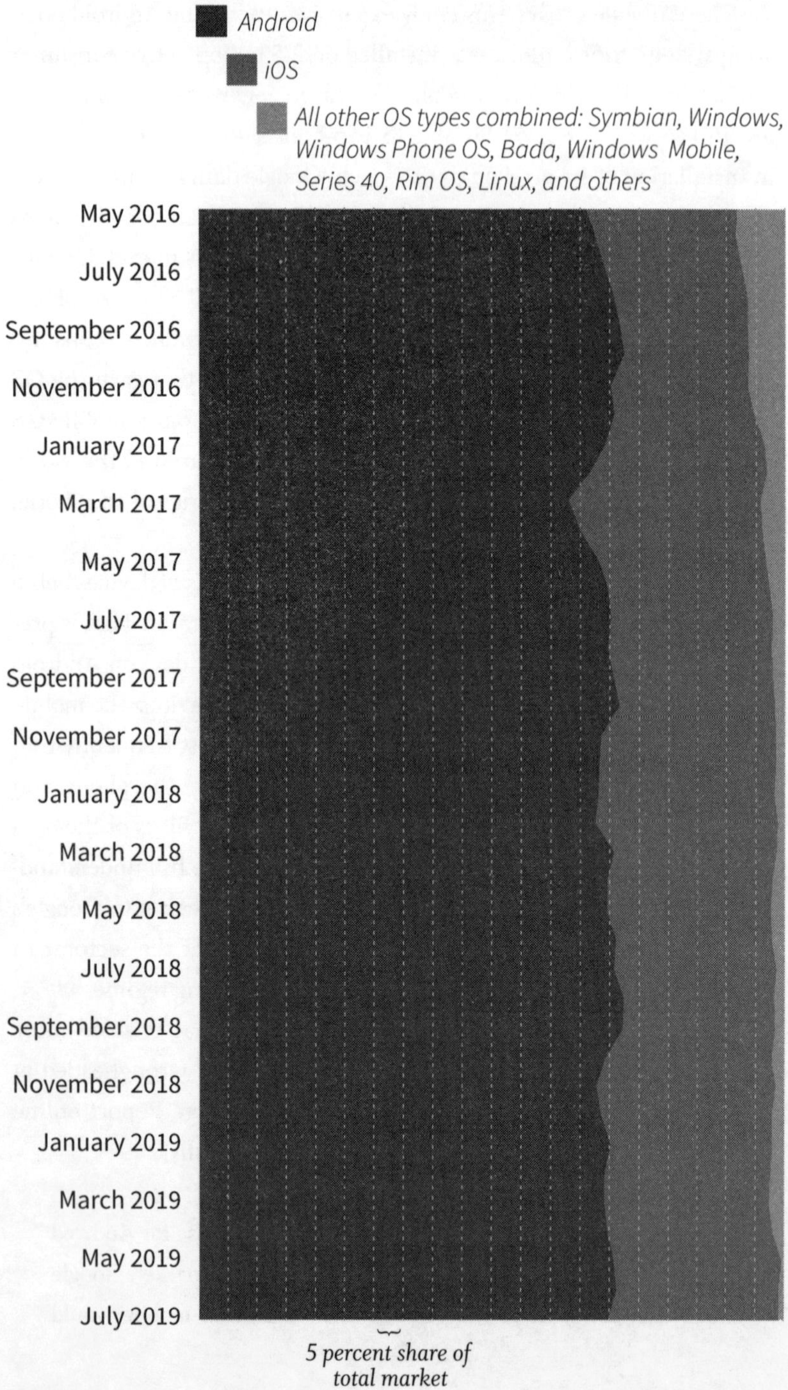

> become more important than PCs, so its search empire needed to thrive elsewhere. Two, that the iPhone needed to be copied. And three, that Android had to be free so that device makers would want to get behind Android as fast as possible, ditching rival platforms like Symbian and Windows Mobile in the process.[79]

In short, making the service free (or at least monetarily costless) was a strategic decision that solely favored Google's commercial interests—not those of the public.

Notably, even companies that do not operate mobile ecosystems might be able to infer similar kinds of information based on your smartphone use. Facebook, for instance, on acquiring Onavo, an Israeli web analytics firm, in 2013, proceeded to spy on smartphone users.[80] Using Onavo, Facebook could gain access to raw data pertaining to mobile web analytics—essentially, the bidirectional flow of commands and data between the end consumer's device and the network operator, since traffic flows through Onavo. While this arrangement might nominally have been used to gain a better understanding of individual users, there was a far more insidious use for the technology: to engage in mass surveillance across the mobile-phone application economy to infer which rival services and products are becoming increasingly popular to consumers and to use those insights to inform the company's overall business strategies and develop competitive responses to rival firms that were gaining traction with consumers.

That competitive response could be instantiated in any number of ways. Depending on the nature of the rival application, its size, and its asking price, Facebook's best response could be to seek an acquisition.[81] If such an option did not appear immediately compelling, Facebook could try to copy the rival[82] or hire away its top talent.[83] Should such efforts fail, it could attempt to make strategic changes to its own product offerings to seal off the market from future competitive threats.[84] (As Ben Thompson of *Stratchery* eloquently puts it, "Instagram's point of differentiation was not features, but rather its network. By making

Instagram Stories identical to Snapchat Stories, Facebook reduced the competition to who had the stronger network, and it worked.")[85] Finally, if all else fails, Facebook could simply drown out the competition through aggressive marketing.

Reportedly, these are all strategies Facebook has executed or attempted in the past. And notably, there are other avenues for ecosystem-level data for technology firms as well—mobile and desktop internet browsers, to name one. Indeed, further evidence reported on by Kevin Roose, Olivia Solon, and Cyrus Farivar has indicated that Facebook—Mark Zuckerberg himself—leveraged user data collected by Facebook to implicitly wield the company's market power in pursuing commercial relationships with other firms involved in the digital ecosystem:

> documents show that behind the scenes, in contrast with Facebook's public statements, the company came up with several ways to require third-party applications to compensate Facebook for access to its users' data, including direct payment, advertising spending and data-sharing arrangements . . . Facebook ultimately decided not to sell the data directly but rather to dole it out to app developers who were considered personal "friends" of Zuckerberg or who spent money on Facebook and shared their own valuable data.[86]

Such overbearing presences in the consumer marketplace breed further exploitation, particularly in a commercial zone where there is little U.S. regulatory enforcement that would otherwise protect the individual consumer and the competitive nature of the market from such questionable business tactics. In fact, these tactics have invited scrutiny from regulators in other jurisdictions, including the European Union, which has argued in a number of cases with respect to several of the leading American technology firms—Google being perhaps the most visible example—that they have engaged in bottlenecking and other anticompetitive tactics to illegitimately stifle would-be rivals and exploit consumer markets.[87] American regulatory agencies, including the Federal

Trade Commission and Justice Department, have also launched serious market competition inquiries in recent times,[88] although some have suggested that the administrative leadership of the relevant enforcement agencies have, under the Trump administration, been overly lenient.[89]

Nevertheless, such regulatory efforts, even if they lead to enforcement actions and apparently hefty fines, do little to break the long-standing and well-established hegemony over data that the leading internet firms enjoy. Even a $5 billion one-time fine against Facebook or Google will only scrape a small margin off profits.[90] That is a small price to pay for forward market consolidation and the open opportunity for increased monopoly power in an internet and technology sector that is steadily more central in our lives as it washes away the more traditional, long-standing entities in the media by undermining their traditional public interest consideration to maintain a level of decency in communication that we were once assured of.[91]

Connecting Boundless Internet Capitalism to the Disinformation Problem

Imagine you are Vladimir Putin's head of propaganda, the lieutenant he has designated to disseminate information—true or false—in the manner that best suits his political objectives. You are sitting in St. Petersburg in front of your staff of 500 programmers and data engineers and just ended a call with Putin's right-hand person: he wants to subvert the American political process again and ensure that President Trump wins reelection.

To you, Putin's motives appear clear. The Americans—and democracy—have seemingly been his sworn enemy since his days as an intelligence agent within the KGB.[92] Everything the United States stands for is anathema to him—and nothing will ever change that.[93] He wants to sow chaos in the American political system; doing so would set the United States back. It would force our nation to ask itself whether the Supreme Court decision to allow gay marriage was the right thing to do.

It would cause us to dither on the question of whether to confer citizenship status to individuals who otherwise would seemingly deserve it. It would engage our political system to enact backward economic policies favoring the industries of the past while ignoring the dire necessities in front of us today. And it would continue to encourage open hatred of the type we have seen in Charleston, Ferguson, El Paso, Dayton, and elsewhere across the country.

But as Putin's lieutenant, just as he wishes to subvert the progress of the West, he also wishes to buttress his own political future. In fact, this is his primary objective: maintain the seat overlooking the Red Square long into the future by claiming that his radical form of governance is precisely what is best for the unique situation of the Russian people. Pointing at America and exposing the faults of its democracy to his political base—in addition to disseminating domestic political propaganda—is an effective means of reminding the Russian people that he is just what they need. After all, look at what has happened in the United States: it is in political chaos, where the boundaries of acceptable civil discourse have shifted so far toward the reactionary extreme that the country appears to be moving backward in time. As Putin's trusted follower, you easily can buy into the argument that the Kremlin needs the authoritative, benevolent, and fair leadership of a man like President Putin; no one else embodies and enforces the ultimate strength of the Russian attitude better. A more democratic approach would only land Russia in the precarious situation the fools across the Atlantic now face.

Your goal must then be to prolong the chaos in the United States and subvert any semblance of a return to order. You have already succeeded in helping to elect Trump. Next comes 2020. Last time around, you did not have a well-trained staff—just a ragtag band of engineers. Now you have a digital army stacked with the smartest young minds in the country, whose singular aim is to collectively support Putin's chief goal during the American election. You have already gathered inordinate amounts of data on the American voting population; thanks to countless data breaches in the United States, you possess enough information to directly or indirectly infer the name, address, political leaning, economic situation, psychological predilection, voting behavior, and likelihood of participation for every one of the more than 225 million potential American voters during the 2020 election. You have set up billions of fake accounts and thousands of shell organizations in the United States over the past four years—a marginal percentage of which has slipped through the cracks despite the best efforts of the internet companies. Even the American government was unable to catch everything. You now possess a direct vector to reach American voters on their smartphone screens. What do you do?

The answer is simple. You take aim at the fault lines—the think cracks—that run through the fabric of American society, those slivers of the population that are disgruntled, unstable, and persuadable. You shower them with targeted disinformation, hateful content, innuendo, conspiracy theories, and vote-suppressing lies that will encourage and trigger their intrigue, their ignorance, and their darker halves.

Keep pounding at those think cracks, and eventually they will break.

The Russians' capacity to exploit our information and systematically push our buttons to the point where our social fabric starts to tear is fueled by the fundamental lack of privacy in this country. We lack the power to find out what information on us these companies have collected and what they have inferred about our personalities. We cannot retract our permissions allowing Facebook to gather up whatever it wants on us through its platforms and the technologies it deploys across the web and in the physical world. We lack a set of fundamental rights—for instance, to opt out of political communications from foreign election interferers and targeted hate speech. We lack the right to privacy.

The Enforcement of Individual Privacy in America Today

The tentacular invasion of the consumer internet industry into every facet of our personhood has corroded the fundamentals of our society. It is the single facet of our economy that may have elevated President

Trump's political chances and eroded our faith in the American political system as an insidious aftershock of the 2016 presidential election. Our personal information—personal identifiers, raw behavioral data, and algorithmic determinations about us—are natural extensions of us. We are the ones who should be in control of information about our personal associations, the media outlets we read, our political leanings, and the businesses we frequent. This is highly sensitive information—it is what makes us who we are. No set of companies with questionable practices should have control over our information—nor should they have the right to use it to control our behaviors, shape our perspectives on the world, and direct us to certain products over others without our clear and explicit permission. We should not be persistently hoodwinked by commercial operators that give no thought to our personal outlooks and desires, despite the apologetic TV ads we might see that suggest they and their machines care deeply about us and the condition of humanity.[94]

No, they do not care. They are profit-seeking machines designed to enrich the coffers of the company owners with cold determination. We have failed to advocate for our rights in the face of this terrible economic logic that subverts the human interest in the name of two things: net income and corporate annual growth rates.

That is the American economic design at work. Silicon Valley—like Wall Street, oil and gas, and the telecommunications sector before it—is an illustration of aggressive capitalistic speculation and inquiry to the most extreme degree. When our national government sets boundaries under a capitalistic design, the most opportunistic businessperson will fill the gap as soon as the opportunity can be seized. That is what has happened in the case of the internet, too—but the boundaries have been set along the wrong lines. They must be pulled back so that they do not tread on our democratic interests. Like a weed that is pushing its way through, we must not only cut off the excited appendages of the industry but attack them at their root.

Some suggest that only now are federal policymakers and politicians discussing the need for privacy reform—and that the discussions about protection of consumers' personal data that have taken place on Capitol Hill are truly novel in this respect. Nothing could be further from the truth. Since the days of Justice Louis Brandeis, particularly given the advent of new communications and imaging technologies since his time, privacy has been a critical issue for various classes of the population.[95] With the coming of the big-data age, however, privacy considerations took a new turn as consumers in the United States and around the world voiced a new demand for fundamental protections guaranteed by the federal government—especially since revelations were disclosed in 2013 about certain surveillance programs operated by the U.S. intelligence community.[96] These demands relate to the desire for individual autonomy in the face of both government and the technology industry, which was revealed to have worked with the U.S. intelligence apparatus.

The Obama administration, understanding these broad concerns, determined that it should initiate a two-pronged policy strategy. First, it was decided that we should work toward the reform of government surveillance programs. Americans and their foreign counterparts had no doubt that surveillance programs in the United States had reached too far and deep into our lives. Second, it was determined that there should be a concerted push to establish meaningful reform—an effort that would necessarily require partnership with Congress. President Obama, recognizing the importance of each of these broad efforts, turned to John Podesta, who had made significant contributions to the development of the Electronic Communications Privacy Act of 1986.[97]

A series of consultations with the key industries that work with personal data, the major U.S. governmental agencies that have a policy stake in privacy and data, civil rights organizations, public-policy think tanks, and consumer advocates resulted in a seminal report on big data that built on the accomplishments of the administration's earlier 2012 Consumer Privacy Bill of Rights. Less than a year later came a robust and much-considered legislative proposal, the Consumer Privacy Bill of Rights Act of 2015.[98] The comprehensive legislation covered the gamut in seven key sections: transparency, individual control, respect for context, focused collection and responsible use, security, access and accuracy, and accountability, all to be backed up by enforcement led by the Federal Trade Commission and the state attorneys general. The president himself announced the legislation during a major cybersecurity summit arranged by the White House at Stanford University and introduced it to his colleagues in Congress early that year.[99]

There were no takers.

The reasons, I believe, are clear. There was limited public sentiment behind the idea of pushing a federal privacy bill at the time, and at an economic level this was largely owing to a combination of hyperbolic discounting, information asymmetry, and industrial obscurantism. Congress was unwilling to break from its inertia of dormancy and do anything about the matter of consumer privacy. Of course, some members of Congress were and remain poised to take on the issue: prime examples are Senators Richard Blumenthal, Ed Markey, Sheldon Whitehouse, and Al Franken (before his ouster), who jointly sponsored a strident bill to force greater transparency on data brokers.[100] But no one budged to take the bill on and introduce it; it was seen as too ponderous and controversial an approach, and as such neither the advocates nor the industry could get behind it.

But the bill, if it had been passed into law, could have accomplished much. Most critically, it would have established an important "baseline." Europeans have long clamored for the United States to adopt a privacy law that would by default afford a fundamental baseline of protections for any kind of personal information. As Daniel Solove writes, we have long-established "sectoral" privacy laws in the United States—sector-specific laws that apply to the field of healthcare, or education, or financial institutions. But this fragmentation of enforcing consumer privacy standards is becoming increasingly problematic.[101] Solove argues that we do not have a consistent approach to privacy protection, which has resulted in a series of problems, including circumstances in which consumers do not know how their data is handled and protected in various sectors. Our laws fail to maintain pace as industries and their use of data evolve over time; major gaps persist in U.S. privacy law, as in the case of behavioral data collected by internet firms; and the international community has no respect for our laws, leading to a multitude of geopolitical concerns, especially with our traditional allies in Europe. We lack default protections that would enforce a fundamental right to privacy, as Europe does. Instead, we are a free market replete with companies that are free to compete in offering privacy protections to consumers. But they do not actually have to offer protections because to date they have determined an approach to public policy and communications that enables the persistent hoodwinking of customers in resolving commercial data privacy concerns. The Consumer Privacy Bill of Rights would have offered a baseline privacy law and could have given us a launching point for more advanced negotiation and discussion. Even if passed into law and later deemed inadequate by the public or policymakers for any reason, it could have been amended as needed. That is, after all, the way that Congress and an administration are meant to work together—not through take-it-or-leave-it political dealings. The quantized, back-and-forth nature of telecommunications regulation in the United States is a parallel illustration of how far congressional and economic regulatory circumstances have regressed.[102]

But the substance of the Consumer Privacy Bill of Rights itself would have helped if it had been passed into law. By affording consumers the agency to work with their data, and giving firms an out from those provisions—a safe harbor—only if they were in compliance with stringent safe-harbor conditions, including the installation of codes of conduct enforceable by the Federal Trade Commission, the provisions of the Consumer Privacy Bill of Rights would have displaced some of the power firms hold implicitly and drawn it back into the consumer's hands. Cameron Kerry and Daniel Weitzner, the former acting secretary of commerce and White House deputy chief technology officer, respectively, and the two leading architects of the Consumer Privacy Bill of Rights, have argued as much.[103] This was in fact part of the reason that the bill was criticized by advocates—because of the safe harbor qualification, which was counted as an excessive relaxation of the requirements that should be applied to firms.[104] Consumer coalitions failed to support the proposal, and it floundered once it reached Congress.[105] But with a reasonable enforcement leadership from the Federal Trade Commission, this regime would have gone a long way toward blunting the disinformation problem.

Perhaps it was simply a different time: the big-data economy was a relatively recent innovation, Facebook and Google had a positive public brand, and we remained hopeful about the prospects of openness and the public nature of the internet—hopes that have since been dashed.

But we are now in a new era, and the privacy situation has regressed. The time for yelling silently has passed. We now must stand up and demand better.

The Temporal Sensitivity of Data

Some have become deeply pessimistic about the internet industry's data-collection practices and that privacy is gone forever. Indeed, the likes of Facebook and Google have engaged in a mass surveillance operation as a matter of business. There seems to be no end to their data harvesting, and there is no incentive for them to delete the personal information they have already accumulated. They have developed sophisticated inferences about us that are seemingly here to stay because they contribute to the high margins experienced across the sector. Furthermore, breaches have occurred: Equifax, Yahoo, MySpace, Target, Facebook, Marriott, Wyndham, the U.S. Office of Personnel Management, the Internal Revenue Service, and others have experienced debilitating unauthorized hacks and exposures of personal information, in most cases to unknown but doubtless nefarious parties.[106] We do not have anything close to any recourse; we lack even a federal law that protects us from this kind of activity. We have fifty state data-breach notification

A CONSUMER INTERNET PRIVACY

BILL *of* RIGHTS

The Obama Administration believes America must apply our timeless privacy values to the new technologies and circumstances of our times. Citizens are entitled to have their personal data handled according to these principles.

Individual Control

Consumers have a right to exercise control over what personal data companies collect from them and how they use it.

Access and Accuracy

Consumers have a right to access and correct personal data in usable formats, in a manner that is appropriate to the sensitivity and risk associated with the data.

Transparency

Consumers have a right to easily understandable and accessible information about privacy and security practices.

Focused Collection

Consumers have a right to reasonable limits on the personal data that companies collect and retain.

Respect for Context

Consumers have a right to expect that companies will collect, use, and disclose personal data in ways that are consistent.

Accountability

Companies should be accountable to enforcement authorities and consumers for adhering to these principles.

Security

Consumers have a right to secure and responsible handling of personal data.

LEARN MORE AT WHITEHOUSE.GOV

The core tenets of the Obama administration's 2012 Consumer Privacy Bill of Rights (a legislative proposal that was introduced in 2015) could potentially have blunted much of the Russian disinformation problem by disabling unencumbered corporate data-gathering, thereby stunting the capacity for micro-targeting.

Source: White House Archives, National Archives and Records Administration, February 2012.

laws in the United States. The financial, telecommunications, and technology industries, however, have fought as a unified front to ensure that a meaningful federal data-breach notification protection that can afford consumers some rights in the face of an invasive private sector will not pass unless it is entirely on their terms.[107]

Given the steady drumbeat of security breaches, some have argued that the degree of data leakage to the private sector and the companies' failure to protect consumer data is so intense that passing a federal privacy law affording access, control, and choice back to the consumer would not reduce the dissemination of private data. Amazon's former chief scientist Andreas Weigend, for instance, has opined, "I have realized that even if you were a privacy zealot, you don't have a chance. Data are being created as we breathe, as we live, and it is too hard a battle to try to live without creating data. And that is a starting point: that you assume that we do live in a post-privacy economy." Even the TV show *South Park* aired an episode in 2014 in which Cartman claims privacy is gone (albeit mostly in reference to drones, which threaten physical privacy).[108]

But this argument ignores an important fact: your behavioral data are temporally sensitive. The data describing your behaviors today are different from the data that described your behaviors a month ago,

The Top Five Largest Data Breaches (2014–2019)

Rank	Company	No. of Users Affected
1	Yahoo	3 billion
2	Aadhaar	1.1 billion
3	Verifications.io	763 million
4	Yahoo	500 million
5	Marriott/Starwood	500 million

Source: UpGuard (www.upguard.com/blog/biggest-data-breaches)

and progressively more so a year ago, two years ago, and ten years ago. Therefore, as behavioral data get older, they become less accurate in describing your true self and therefore less valuable to the internet companies. Back when you were in college you might have had entirely different tastes. Those tastes determined your spending habits in the market. It is the real-time, currently relative knowledge of those habits that matters most to companies like Amazon—not information about your past behaviors. If you just watched the ending credits of the *Curb Your Enthusiasm* series, maybe you would consider buying the entire Blu-Ray disc set next, or maybe you would be interested in purchasing Larry David's white sneakers or watching *Seinfeld* on Hulu or YouTube next. If you were spotted yesterday checking out the health food aisles at Whole Foods and gym shorts at Lululemon, then marketers at Nike, SoulCycle, Equinox, Spotify, and Brooklyn Boulders should automatically be reaching out to you to see if you might be interested in signing up for their related services.

It is less interesting for these companies to know that you were spotted in these aisles a year ago; your interests might have changed, you might have had to take on a more demanding job, or you might have skipped town altogether. As such, your purchasing behaviors in the near future do not reflect your spending a year ago as accurately as your spending yesterday does. This temporal disparity is precisely why information gathered on you recently is given far greater mathematical weight by marketers than older data about your behaviors. This is precisely why firms such as Nielsen, in its current evolution, is essentially the world's most established data broker, designing technologies "for the collection of market research data [on a real-time basis] from a plurality of cooperating retail stores, each of which utilizes Point-Of-Sale (POS) optical scanners/registers and associated automatic controllers . . . [more specifically concerning] a substantially [and] totally passive or noninvasive automated system"—or, in plainer terms, technologies that attempt to infer real-time behaviors seamlessly through analysis of consumer actions in the marketplace.[109] There is a whole field in marketing science behind real-time insight development called sentiment analysis, which modern firms practice using advanced machine-learning techniques.[110]

This analysis indicates one thing: if we could pass a meaningful privacy law today, it would have an immediate effect on your life. You could opt into or out of these kinds of data-collection programs and the inferences built atop them. And as such, overnight you could have tremendously more power in the face of the digital behemoths.

The Future of Privacy: On the Edge of Innovation

We live in a cold new world in which the modus operandi of the industry is to collect as much data as it can. This is largely because the marginal technological cost of collecting it is minimal; we now have television shows that poke fun at the ease of setting up a server in your apartment's storage room and running a fledgling start-up out of it.[111] Meanwhile, the asymmetry of knowledge about the practice of data collection, analysis, and monetization has left consumers well behind the industry. We are, every one of us, economically exploited; the industry's goal is to enter our mind and move our psychology to the point where our market actions become influenced in their commercial favor. The most striking feature of this terrible new circumstance is that this algorithmic machine has matured organically. It is not the result of a concerted business plan but rather an experimentally and empirically evolved animal trained to identify opportunities for economic arbitrage in the novel industry of manipulated communication.

There might be technical solutions that can yet be developed to overturn this new regime of mind control. There is a burgeoning academic field—called privacy engineering—that is forging new advances in the technical protection of privacy and security. Some fascinating applications are beginning to emerge out of this field. Federated learning, by which machine-learning algorithms are deployed over distributed networks of devices rather than centralized at, say, a service operator;[112] differential privacy, which puts forth a mathematically rigorous definition of privacy and offers a set of noise-injection and data-aggregation algorithms that if implemented can preserve individual privacy;[113] and fully homomorphic encryption, which enables computing with encrypted data without needing to decrypt the data[114]—all have been developed in recent years and have added new richness to the development of privacy-preserving technologies.

These represent real mathematical progress; if advanced and eventually implemented, they would change the privacy situation through technology without the need for regulatory reform. But there is a fallacy in industrial incentives. Research I conducted with collaborators showed that in the absence of regulation that forces the industry to adopt the privacy-preserving solution, the incentive is for the industry only to continue down its current path. There is no market for privacy because consumers fail to collectively demand it in the United States.[115]

In light of unbalanced and uninformed consumer markets and the deviousness of an internet industry that inherently exploits the citizen's lack of understanding, we will need to artificially create the market for privacy, and there is only one way to accomplish this: informed regulation on the back of citizen engagement calling for such governmental intervention.

As we move forward, though, privacy will take on more complicated dimensions. We have analyzed the internet privacy situation here through the traditional lens, assessing the legitimacy of data collection in and of itself. As the industry silently moves more aggressively into the world of ubiquitous data collection, processing, and analysis, commercial entities will make decisions that pertain to our lives simply to turn a profit with increasing frequency. If those decisions—over what media we should see, whether we should get a loan to buy a car, what kinds of job postings we should be shown—were made fairly and exactly according to our desires, I doubt we would care that some companies were making money off our data by attempting to infer what we wish to see.

But as we will discuss in the next chapter, the fact is that we have become the willing fuel for a corporate machine designed only to cater to shareholders at the expense of citizens. We are experiencing the height of the era of "commercialized decisionmaking."

THREE

Algorithms

The Commercialization of Bias

The importance of the consumer internet in the modern media ecosystem is beyond question. Economic opportunities in housing, employment, and other dimensions of the consumer marketplace,[1] national political concerns and the systemized dissemination of political communications,[2] and social interactions that mirror or define our sociocultural norms[3]—these are all clear and evident results of the growth and present breadth of influence of the consumer internet.

The consumer internet is made up of the firms that operate over the internet and interface directly with consumers—Facebook, Apple, Google, Netflix, Spotify, and Amazon among them. Consistent across the sector is a set of practices that are characterized by three elements: engaging platforms, behavioral profiles generated by the collection of the consumer's personal information, and highly sophisticated algorithms that curate content.

Two caveats are in order before moving forward. First, while this business model is in clear use within the walled gardens of such firms as Facebook and Amazon, each firm adopts the model in its own way to take advantage of the profits it can yield, evolving its singular economic logic through proprietary practices and with the goal of collecting personal data, along with its unique value proposition for the consumer market. Second, this business model is used to varying degrees by individual firms; there may be other core practices and contributions to company revenue that are also critical to the company and are operated in parallel with the consumer internet offering. For example, Amazon and Google are market leaders in the provision of cloud-computing services; Apple's core revenue is generated from the sale of consumer-device technologies; and Netflix maintains an order-by-mail DVD rental service that has relatively little to do with this business model, leaving aside the agglomerations of personal interests derivable from physical rentals.

The practices leading to the business model illustrated above constitute what I define as the "consumer internet," and it is these that need to be scrutinized and critiqued. This is the business model that has instigated and perpetuated the negative externalities inflicted upon the public today, precisely because it has affirmed an alignment in the persuasive interests of nefarious actors and the economic interests of internet platforms.

Why have these firms uniformly adopted this business model? The internet industry operates in a free commercial zone—it is, in other words, a radically free market that favors and rewards open capitalism.[4] The United States lacks a federal standard on privacy and most other public-interest concerns that would otherwise govern the firms in this sector.[5] This fundamental absence of protection of American consumer and citizen rights has given the consumer internet firms a free pass to take advantage of the free market zone. And take advantage they have, as suggested by the institutional directive within Facebook to "move fast and break things."[6]

The absence of a regulatory regime has meant that these firms have developed in a manner that is practically independent and that disregards the public interest, save when it serves their commercial interests. As is the case for any American business, the public interest need not be considered; only the shareholders' interest need be served. Chief executives in the industry do not commonly bow to consumers. In the absence of meaningful economic regulations that target the capitalistic overreaches of the business model, only market forces themselves can save the public from the industry's intrusion into the personal lives of consumers. And market forces will not do the job; the market has lost its fluidity and efficiency, thanks to the economic stranglehold that the firms at the top of the internet industry have had. This is perfectly illustrated through Facebook's mimicking of the most imaginative features developed by Foursquare, Path, Tumblr, and Twitter, each of which once believed it might have a path to true competition with the internet behemoths that preceded them but thereafter substantially faded, albeit to varying degrees. Such carbon-copying of actual innovation was only possible because Facebook had inordinate power over the social media industry and the financial and technical wherewithal to swiftly subsume its smaller counterparts and unceremoniously absorb their most attractive functionalities.[7]

While public sentiment may at times swell to such a degree that it might appear an effective impetus for firms to change, the total standstill in meaningful regulatory progress by the U.S. government enables the free zone of digital commerce exploited by Facebook and Google to remain intact. The public's memory is short, and once the furor has quieted, the industry often returns to its normal commercial operation, perhaps under new disguises to protect itself from the regulatory community.

This might be precisely what has happened in response to the disclosures of March 2018 that Cambridge Analytica had shared the personal information of 87 million Facebook users with the digital strategy firm engaged by Donald Trump's campaign for the American presidency. While public outrage immediately followed the whistle-blower's revelations and the corresponding reports—outrage that led Facebook's chief executive Mark Zuckerberg to testify before Congress mere weeks after the revelations—there is relatively little U.S.-led discussion now about what economic regulations should be passed to hold Facebook and like firms to account.[8] Despite congressional hearings,[9] media statements,[10] white papers,[11] legislative principles,[12] and bill introductions,[13] there is relatively little understanding of the industry or, for that matter, any clarity on what should or even could be pursued by Congress to promote much-needed transparency in the digital political advertising hosted by internet firms.[14] Now with Mark Zuckerberg's commercially convenient commitment to free speech and Facebook's corresponding shirking of all duty to remove ads that a politician can use to spread lies and disinformation, online political advertising has become a true new Wild West.[15] When the president of the United States can disseminate the bald-faced lie that perhaps his greatest Democratic political rival, former Vice President Joe Biden, exchanged promises of government money for personal favors,[16] we know that our democracy and the commercialism it supports has reached a new midnight, a deeply dark corner. We must bring ourselves out from it.

It is this unrestrained reach of the business model—particularly the constant quest by firms like Facebook and Google to maximize, through whatever means necessary and possible, the amount of time users spend on the platforms—that has allowed the leading internet platforms to overrun political communications and media in America.

The Science of Machine Bias

Robust scholarship has developed in recent years, particularly since the boom of the big-data economy, concerning the potential of machine-learning algorithms to systematically perpetuate discriminatory results in various fields from medical science to educational opportunities.[17] Of primary concern is the development of machine-learning models that engage in automated decisionmaking. While the methods underlying the application of machine learning are mostly taken from the traditional statistical literature, cultural circumstances and advances in computing have popularized the term and expanded interest in the industry.

"Supervised" machine-learning models—to which there are many underlying algorithmic approaches—are typically designed through a combination of human input and automated statistical analysis of a data set.[18] A data set—such as the demographic inferences Google draws about a class of users in Manhattan—typically carries some implicit pattern: different classes of users might execute searches at various times of the day, from various locations, and with varying frequencies—indicators called "features," since they are independent attributes associated with an instance, in this case an individual user. Machine-learning models attempt to draw such relationships out of the data to develop inferences about the true nature of the population. Supervised machine-learning models have been applied by researchers and industry engineers to seemingly every data set imaginable—from clinical measurements of cardiac arrhythmia[19] to forest trails for mountain robots.[20] Engineers are even developing applications to synchronize broadcasts independent of geography so that when LeBron James pulls up against Kevin Durant, NBA fans in Los Angeles and Brooklyn will get to watch the ball leave his fingertips at exactly the same moment.[21]

In the consumer internet industry, a human—or, in the case of unsupervised models, a machine—might code (or categorize) each user of a platform as a participant in a particular class based on the user's individual features, as inferred through analysis of the user's on-platform behavior, off-platform activity, and demographic data. A set of "training data" is thereby generated that can be used to help the model learn how to classify future data points. A data set that includes a population of such users might have some observable relationships consistent across certain classes of the population (for instance, that kids of a certain age are more associated with purchasing toys or basketball trading cards). These relationships are drawn into the machine-learning model in the form of a set of decision rules—a series of inferences about the population developed from observation of the data set that can be implemented as a classifier of future objects, subject to the model's classification regime. This implementation can then be executed on an automated basis such that when new observations come into view, they can readily be analyzed and classified by the model.[22] Machine-learning models designed to infer behavioral profiles and curate content are refined by consumer internet firms through feedback from real-world routines and behaviors.

We can regard YouTube's video recommendation system, which suggests what video the user might watch next, as the model or classifier operating over a set of decision rules established by the machine-learning model developed and routinely refined by Google. The company's commercial objective is tied to engaging the user, thereby enabling YouTube to collect increasing amounts of information about the user's habits and preferences, as well as to generate ad space that it can sell to the highest bidder interested in targeting a set of users.[23] Particularly at the outset of algorithmic design, a team of humans might be employed to classify a global set of users into various categories for each feature. For instance, a feature concerning type of use of the platform might include classes such as channel operators, power users, frequent users, and occasional users. Additional features might include demographic details; information pertaining to the user's historical use of the platform, including which videos the user watches and which channels he or she has subscribed to; information pertaining to the use of other Google services; and position in the relationship graph network. YouTube might then train a model to analyze how the existing observed data points concerning the company's users were classified.

This analysis of observed data points is used to develop, train, and routinely modify a set of decision rules constituting the classifier model that can determine, based on statistical analysis, which class the new data points—new users of YouTube, in our example—should be entered into. Finally, this algorithmic inference determines what videos the platform will recommend to the user. Feedback loops incorporating

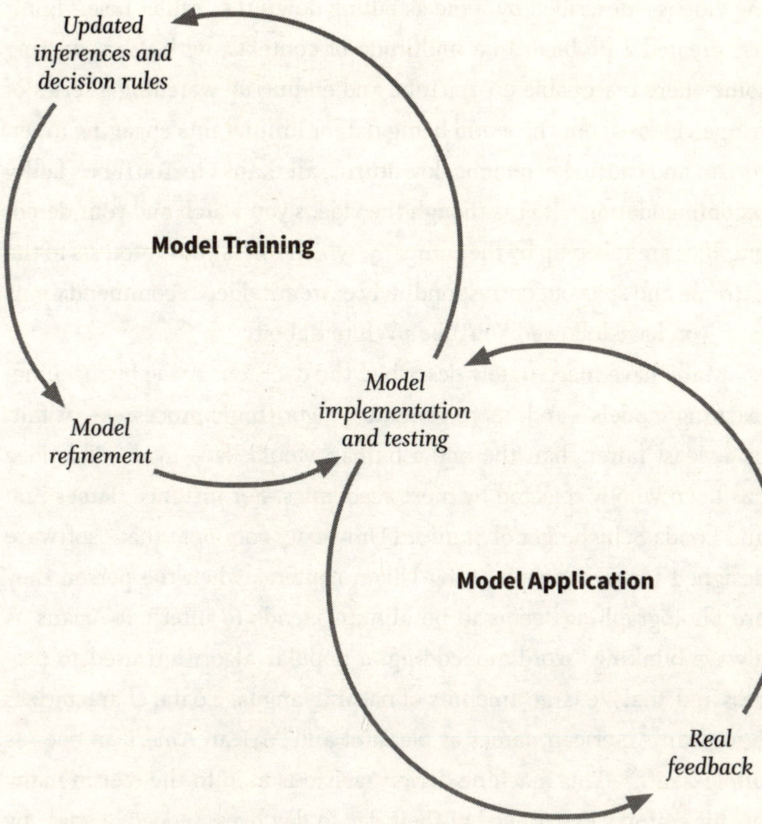

Commercial learning models designed to infer behavioral profiles and curate content are continually refined by consumer internet firms through feedback from real-world routines and behaviors.

accuracy of predictions (whether reported by the user or inferred by the platform in response to a user's disengagement or other negative behavior) can be used to refine the model over time. This, in turn, may lead many users down a path of watching a long series of highly engaging videos—described by some as falling down the "rabbit hole." But it has created a problem in a multitude of contexts, with users starting somewhere reasonable on YouTube and ending up watching a series of fringe videos about the world being flat, or immigrants engaging in terrorism and causing economic downturns, all thanks to YouTube's faulty recommendations. It is as though the videos you watch and your demographics are taken up by the company, which infers your interests to the extreme and spits out correspondingly extreme video recommendations once you have followed YouTube's White Rabbit.

Many have inaccurately described the decisions made by machine-learning models—and, more generally, algorithmic processes—as fair, or at least fairer than the one a human would have made. This idea has been wholly rejected by most academics. For instance, James Zou and Londa Schiebinger of Stanford University comment that "software designed to warn people using Nikon cameras when the person they are photographing seems to be blinking tends to interpret Asians as always blinking. Word embedding, a popular algorithm used to process and analyze large amounts of natural-language data, characterizes European American names as pleasant and African American ones as unpleasant."[24] This machine-driven racism is akin to the racism many in this country experience in their day-to-day lives. Indeed, a study by Marianne Bertrand and Sendhil Mullainathan illustrates the prevalence of such parallel racism among résumé readers who favored names associated with certain races over others.[25]

In the application of learning algorithms, as in all things, we nominally wish to execute only those decisions that are just—and as John Rawls famously argued, one must see "justice as fairness: that participants in a common practice be regarded as having an original and equal liberty and that their common practices be considered unjust unless they accord with principles which persons so circumstanced and related could freely acknowledge before one another, and so could accept as fair."[26] Focusing on the idea of fairness, we can note that while theoretically algorithms could be designed in a manner that is contextually fair, one question that arises is what fairness (even in context) should actually mean. Fairness is a philosophical concept that has no agreed-on definition, and as the philosopher Jonathan Wolff has argued, contemporary egalitarian theory has yet to provide a clear answer as to how society should approach the enforcement of fairness.[27]

This has been a particular concern in the business community for many years, as companies have tried to determine what constitutes fairness so that they can calibrate their service terms to norms that consumers will find acceptable—particularly since customers who think they have been wronged will resent the company.[28] A paper by Daniel Kahneman, Jack Knetsch, and Richard Thaler addresses this issue, identifying phenomena where perceptions of fairness drive the behaviors of consumers, tenants, and employees—classes of individuals who are often economically dominated and exploited by the firm.[29] A further complication arises when different parties might have different definitions of fairness in practice. Even where there is broad consensus on the meaning of fairness, machine-learning models have been shown to discriminate, as in the case of Google Photos's image recognition technology, which labelled two black people "gorillas."[30] (Chillingly, a company as capable as Google only "fixed its racist algorithm by removing gorillas from its image-labeling tech.")[31] Yet another concern is that it has proven to be difficult designing machine-learning algorithms in a manner that foresees all potential forms of unfairness and preempts them through intelligent reorientation of the algorithm.[32]

In the consumer internet context, should "fair" mean "lawful"—and, correspondingly, should everything outside the reach of the law be permissible and therefore fair? That is essentially today how the industry operates—and it is the underlying free market economic design of the United States that enables and encourages such capitalistic "innovations" as discriminatory decisionmaking executed by artificial intelligence so long as it does not constitute harmful disparate impact in a manner that violates federal civil rights laws. In this way, the vast majority of the consumer internet's industrial activity falls directly outside the purview of federal antidiscrimination laws in the United States. Sometimes certain advertising practices might even clearly appear to be egregiously discriminatory, as a group of researchers found with Facebook's ad-targeting mechanisms, which exclude users belonging to a certain race or gender from receiving their ads.[33] But there is seemingly little that can currently be done to prevent discrimination in online targeted advertising. Only when the business activity concerns American civil rights laws—as has been suggested by the American Civil Liberties Union about a very small minority of Facebook's advertising practices—do consumers have a leg to stand on.[34]

A Predilection to Segment Society into Classes and Discriminate

Machine-learning models discriminate. This is precisely what they are meant to do: discern the characteristics of an incoming data point and infer, based on its features, which class it belongs to. Presumably, such models used give potentially different treatment to data points that occur in different classes. In the case of YouTube, *Game of Thrones* fans might consistently be recommended videos related to conspiracy theories concerning the future of Westeros and what Arya Stark will encounter on her ventures west; car enthusiasts might be pushed toward reviews of the new Ferrari 488 or highlights of the latest Formula 1 race, or if they recently watched videos about how to install wider racing wheels, videos about Yokohama tires and carbon fiber spoilers. But problems arise as these platform algorithms caricature our interest profiles and extrapolate them to the extreme. As Guillaume Chaslot, an early YouTube search-algorithm architect, has put it, the "algorithm for the last 10 years has been pushing people down rabbit holes," necessarily so because "that is most efficient for watch times."[35]

Title VII of the Civil Rights Act offers employees protection from unfair decisions made on the basis of various protected categories, including race, gender, pregnancy, religion, creed, veteran status, genetic-testing status, ancestry, and national origin.[36] (Protection from political discrimination is not included.) Various laws institutionalize further protections, among them the Age Discrimination and Employment Act, which stipulates that employees cannot be terminated simply because of their age; there must be some substantiation that the aging employee no longer can work effectively.[37] Similarly, the Americans with Disabilities Act prohibits employer discrimination against individuals who can work effectively despite their disabilities.[38] Various state laws go further, instituting additional protections from discrimination, particularly by adding protected classes and other expansions, including, for instance, lower age thresholds to trigger the age discrimination law. A progressive state like Connecticut, for example, prohibits discrimination on the basis of any age—young or old.[39]

Developing civil rights jurisprudence protects against two principal mechanisms of discrimination: disparate treatment and disparate impact.[40] In a typical disparate-treatment case, a potential employer might suggest that the candidate will not be hired because he or she is the member of a protected category. This sort of determination would amount to an intentional violating decision to discriminate against the candidate because of his or her protected class status.

But in the realm of machine bias, it is the disparate impact cases that are typically of greater concern because of the manner in which learning algorithms engage in automated classifications over which decisions—decisions that could be vitally important to the data subject in question—are automatically applied and implemented against many data subjects together, according to a set of rules contained in the model. Disparate impact cases typically refer to instances in which a particular decision has greater resulting impact on a protected group than on the rest of the population. Harmful disparate impact can trigger an investigation against the liable party. And a decision such as a hiring policy might be "facially neutral"—where the decision rule does not appear to be discriminatory on its face—but if in practice it results in a harmful disparate impact against a protected group, civil rights protections may be triggered.[41]

A learning model that classifies users in a consumer internet application—for instance, in identifying the consumer group at which to target a set of political ads—might attempt to maximize click-through rates or some other engagement or revenue metric applied by the platform firm. The learning model might identify characteristics regarding a number of signals (or features) about an advertiser's messaging and intended target audience—for instance, ad content featuring men and masculine themes, certain socioeconomic or cultural themes (such as eau de cologne, Swiss watches, and deluxe shaving equipment), and the Northeastern geographic region. It is likely that the algorithm would then determine that the target audience that will yield the greatest engagement for the advertiser and the platform is some group that is correspondingly primarily male, wealthy, and Northeastern—which, it could be said, is necessarily a harmful and discriminatory targeting practice, given that certain protected classes are not included in the target audience.

That said, such targeting is probably not illegal for several reasons. First and foremost, there might be no civil rights law that covers the content of the ad campaign in question. American laws primarily cover various economic opportunities but not social or political contexts. Second and perhaps more critically, it might be that even if the ad content is covered by civil rights laws and pursues a discriminatory execution of dissemination that prevents certain protected classes from seeing the ad, the classifier was technically "fair" per our legal regime. In such cases, if a suit is pursued, the platform firm that enabled the targeting may have to respond to the question of why the algorithm screened out an inordinate proportion of, say, a certain race. Should the firm be able to offer a justifiable business reason—for example, that it was only targeting past established customers for whom it possessed email addresses—then under our current legal regime it would most likely be adjudged not guilty of having engaged in unfair discriminatory practices leading to harmful disparate impact.[42]

The Discriminatory Concerns of Training Data

Broadly speaking, the use of learning models can produce discriminatory outcomes through two main means: the nature of the training or input data and the design of the learning algorithms. Underlying each of these themes is a more human concern: that data miners themselves could be (intentionally or unintentionally) biased and could carry that bias into the programming of the model and analysis of the data.

There is a long-running refrain in the field of computer science: "Garbage in, garbage out."[43] Machine-learning models are "trained" through the analysis of training data, which, in supervised learning schemes, might be classified by humans. Any given data point—such as a typical Snapchat user—has a set of attributes about his or her use of the company's platforms that can be used to classify the user into certain audience clusters. Inferences about new users to the company's systems are then made by the learning model. But, as discussed in a May 2016 White House report, poor design of training data can promote discriminatory outcomes.[44]

Flaws in training data can perpetuate discriminatory decisionmaking by two primary mechanisms. The first originates in the process by which the data are organized. Historical data sets on which training data are based typically come with certain mutually exclusive class fields, as discussed earlier in the YouTube example; but the selection of class fields and attribution of data subjects to them occurs at the hands of humans in supervised learning premises. The people who organize these class fields—the data engineers responsible for development of learning models—attempt to define a paradigm through the identification of class fields that they believe most fairly and effectively reflect the situation of the real world.[45]

For instance, it might be that to make determinations about the creditworthiness of a loan applicant, credit agencies decide that it is most critical to understand his or her net worth, demographic information, profession, education level, and related categories but that to include information related to the individual's personal life goals, trustworthiness, and commitment to paying back the loan is less important or is particularly difficult to ascertain. This can germinate a form of bias in the designation of class fields, as determinations to include and exclude certain categories could diminish the chances of a positive decision for certain demographic groups while elevating the chances of others. The creditworthiness example can be translated to the consumer internet context: firms continually refine ad-targeting algorithms so as to advance the commercial interests of the advertisers by offering them the maximized bang for the buck with the data that they have at hand. Whether the advertiser is a credit, housing, or employment agency or another client, the tendency for all parties involved will be to promote profits over protecting the consumer's interest, given the lack of any sort of legitimate nonpartisan scrutiny over firms in the digital advertising and consumer internet sectors. While certain statistical sampling techniques can be applied to resolve this issue, this necessitates an extra step that most service providers would quite likely find unappealing were they not publicly scrutinized or forced to undertake as a matter of compliance.[46]

The second major family of discrimination concerns that might arise from poor design of training data sets is attributable to the data themselves. Two main problems can be responsible for this: inaccurate data and selection bias.[47] In the first case, data might be outdated or otherwise contain inaccuracies about the population that perpetuate bias since the incorrect data are used to train the classifier model. For instance, if loan payback periods are incorrectly reported to be longer for some individuals than for others, then those individuals might be adversely affected by the decisions executed by the resulting model trained on the inaccurate data set. The second case, selection bias, is often subtler and involves the collection of data that are not representative of the population, which, if used to train the resulting learning model, projects the inferences learned from the biased training set on current decisions, resulting in likely biased decisions.

A simple example of biased input data occurred in the case of the StreetBump application developed in Boston. The mobile application was designed to enable residents to report the occurrence of potholes to the app developer. The idea was seen as so successful in enabling crowdsourced reports that the municipality engaged the developers to inform them when and where to dispatch repair teams. After some time of use, however, it was found that repair teams were disproportionately dispatched to wealthier and younger neighborhoods—parts of the city that presumably had more people who owned smartphones and greater local propensities to participate in the crowdsourcing functionalities offered through the application. The city was, in other words, receiving a biased selection of the data. A truly representative set of data would report relative frequencies of potholes across neighborhoods in the city in proportion to their true occurrence. Thus without using a more representative sampling of data or tweaking the algorithm to correct for the direct harms that came to the neighborhoods that were less well off, the process of deciding when to dispatch repair service could only produce biased results.[48]

A related concern is the capacity for learning models to suggest discriminatory decisions based on biased data sets. Training data might only contain information at a level of granularity that disadvantages certain groups. The issues around the granularity of the data sets in question lead to such potentially discriminatory practices as redlining, in which certain inferences are drawn about individual neighborhoods—inferences that advise decisions made about any residents living in those neighborhoods. If the data suggest that average earnings in a certain zip code are relatively low, the inference could be that users in that area will be disinclined to click through on ads and eventual purchases of interesting market opportunities. Thus anyone living in that neighborhood could be subject to a discriminatory outcome that may constitute a harmful disparate impact upheld by the courts should the harm occur in regard to, for instance, a housing opportunity.

Optimal Algorithmic Design and the Industrial Pursuit of Formulating Bias

Machine-learning algorithms carry the bias contained in data inputs and reflect those biases, as the model learns based on the makeup of the training data. But there are additional concerns that can result from the mechanics of traditional statistical analysis as well.

Foremost is the common fallacy in statistical analysis that correlation necessarily implies causation. We know this not to be true.[49] Educational levels might vary across certain racial groups, but this should not be taken to suggest that certain races are more intelligent or hard working than others, contrary to what Charles Murray infamously argued twenty-five years ago in *The Bell Curve*.[50] Though this concern has resurfaced with machine-learning models, there are mechanisms to curtail its prevalence proactively, in much the same way that certain explanatory variables are excluded from regression models because they are redundant or misleading.

Perhaps more deeply concerning, a poorly designed machine-learning model—or one that is ill equipped to fully handle the problems of discrimination, especially in areas that are not subject to strict regulations, such as personal finance and housing—may drift over time in such a way that biased outcomes for marginalized people are perpetuated.[51] This problem is distinct from the initial training of a model; indeed, trained models implemented in the consumer internet industry are refined on a regular basis so that they reflect the user's desires to the greatest degree possible. But what happens when an algorithm exceeds its intended purview and presumes things about us as individuals or as a population that are just not true—or even worse, encourages engagement of our less virtuous tendencies? A widely known statistical concept describes a related tendency: confirmation bias, whereby the model or its designer finds what might be expected, given cultural norms, instead of the reality of the world. As psychologist Raymond Nickerson puts it, confirmation bias is the "seeking or interpreting of evidence in ways that are partial to existing beliefs, expectations, or a hypothesis in hand."[52]

The broad propensity for learning algorithms to drift presents a thicket of concerns regarding bias. For instance, a model might learn from original training data that have been carefully engineered and monitored by the data miner to limit occurrences of unfair discrimination. But at their center, learning models are designed to cut corners, to efficiently make decisions and determinations about a population in a way that approximates the true nature of the real world and reflects that in its algorithmic design—and as such, they are designed to discriminate. There is no avoiding this impact: most learning algorithms in use today cannot help themselves because they are designed to optimize within a given set of constraints, and their fundamental tendency will always be to push out to the absolute edges of what is considered fair.[53] That is precisely why companies are driven into corners, as Amazon was when it was discovered that its artificial-intelligence recruiting technology was biased against hiring women.[54] However, the fact that Amazon shut the program down should not necessarily make us feel that antidiscrimination efforts are working. Amazon's oversight of their hiring practices was particularly sensitive because there are strict civil rights laws in place concerning employment, and the hiring decisions pertained to Amazon itself—so the company likely had to act assertively to limit reputational damage and, potentially, lawsuits.

But such stringent regimes as federal employment antidiscrimination laws do not exist for the broader set of instances of silent discrimination that we are concerned about in the consumer internet forum.

This natural tendency for machine-learning models to attempt to find ways to discriminate in whatever legal manner possible inherently forces them to tend toward overstepping the boundaries that have been set for the industry through secondary back doors, which, in turn, enforces within the decisionmaking model an economic logic that drives it to acquire new behaviors through novel discoveries about the real world. Indeed, this is precisely the way that Facebook's news feed, the Twitter feed, YouTube's recommendation algorithm, and countless other consumer internet suggestion systems work. But what happens when the so-called "discoveries" that advise the decisionmaking algorithm are outsized or otherwise biased?

This is the type of model drift—through ongoing observation of the real world—that can engender discriminatory behavior. It is this characteristic of machine learning that can cause models to systematically feed voter suppression content to marginalized groups,[55] or send nationalistic groups down hateful-content pathways,[56] or implement racist crime prediction software that especially harms African Americans across the United States.[57]

The systemic generation of "proxies" as the model is trained from the input data can also subvert antidiscrimination efforts. Anupam Datta and his colleagues define a proxy as a "feature correlated with a protected class whose use in a decision procedure can result in indirect discrimination."[58] It may be that a learning model is designed to exclude the use of any protected-class data in the course of statistical analysis so as to explicitly protect against discriminatory outcomes concerning those protected classes. Models might learn, however, that there are alternative proxies that are equivalently descriptive of the protected-class categorization as the protected-class data themselves. For instance, an algorithm prevented from accessing race information might determine that some combination of other class fields—such as location of residence and name—might be used in tandem to generate through the back door an understanding of an individual's racial group category. Furthermore, such inferences might be completely nontransparent to the model's engineers, since they typically occur silently, premised on the data already provided as input to the model, and proxies are not proactively reported to the designers, as they are generated by the learning algorithm in the course of maintaining and updating the classifier model.

Sometimes proxies can be explicit. Facebook, for instance, designed three primary advertising-interests categories known as "ethnic affinity" groups for the U.S. market: African American, Hispanic, and Asian American.[59] In other words, advertisers could choose to target people who Facebook inferred have some interest in Asian American cultural themes. This was most likely done to obviate the need to use actual race data, which could have prompted discriminatory concerns and thereby triggered class lawsuits off the bat. In practice, not every African American user would be in the African American ethnic affinity group, and not every member of the African American ethnic affinity group would be African American—but that did not change the behavior of Facebook's advertising clients, who have found these interest categories highly lucrative. This enabled an insidious discriminatory feature in machine learning that requires correction. For example, in 2016, ProPublica managed to purchase and create an ad for people who might be househunting and excluded users with "an affinity for African American, Asian American, or Hispanic people. . . . Facebook declined to answer questions about why our housing-categories ad excluding minority groups was approved 15 minutes after we placed the order.[60]

Whether most of the members of these groups are members of the corresponding race—and whether therefore each affinity group represents an explicit proxy for the relevant race—is an open question. One could argue that the correct design and designations behind these groups are even more useful to Facebook and its advertising clients than factual race data: knowing that someone is interested in—and therefore probably willing to spend money on—products that Facebook associates with African American culture represents an even more direct signifier of intent than the actual race data themselves—especially as racial stereotypes might not hold for every member of a race. It may even be a little more questionable in legal terms, too. Advocates have long battled the industry on this front,[61] and perhaps the situation will change over the coming years, should advocacy groups continue pushing with the force of government to back them up, as we have seen in the Department of Housing and Urban Development's leadership in recent times.[62]

The harmful effects of discrimination in machine learning nevertheless become supercharged when it is in the direct commercial interest of the party developing the learning model to develop a classifier that maximizes revenue. In such an environment, potential discriminatory outcomes become a mere afterthought to serve the needs of corporate compliance. The cutting edge of the model goes undisturbed, and like a hot knife through butter, it finds ways to slice society into slivers, all to the end of making a (highly profitable) buck.

The Radical Commercialization of Decisionmaking

Is the fact that internet firms have overtaken the Western media ecosystem actually bad for our society? Perhaps it is for the best, in that it breaks the more traditional centralization of content creation?

A truly social platform should elevate not just the content generated by actors in the mainstream media, as major newspapers do, but rather those issues and elements raised, reported, and reposted by the common user, and particularly a mix of those posts that are predicted to be interesting to the audience user in question and that have also received wide circulation. (Other factors are at play, including the explicitly expressed preferences of the user, who might, for instance, choose to view his or her social feed in chronological order, obviating some of the concerns recently associated with social media platforms—though on Facebook's current news feed this has to be done every time a user logs in to the web service. A cynic might suggest that Facebook does this for reasons besides its oft-trumpeted claim—which is likely true—that when polled by the company, users indicated that they tend to prefer Facebook's manipulated ranking over the chronological ordering.[63])

Thus the centralized power of large media companies of years past—epitomized by the Hearst Corporation, among other examples of the twentieth century, which was in part responsible for triggering the Spanish-American War during an infamous era of yellow journalism[64]—is diminished by the nature of the internet and the internet platforms themselves, particularly as social media receives more attention from younger generations than traditional forms of print media that offer a different kind of access to the news. In fact, most of the appeal of social media originates from its capacity to connect us with issues and ideas that matter in our individual lives—issues that would not at all appear on traditional media formats—more so than the more abstracted concerns of the mainstream media. Neither William Randolph Hearst's papers—nor the modern *New York Times*—were any place to disseminate one's personal reviews on the quality of Laphroaig 18 Year Old Scotch Whisky. Indeed, the distillery turned to digital media to engage in a campaign to solicit perspectives on the single malt's unique taste.[65]

Whereas power among the producers of the traditional news media has waned, the power of the internet platforms has quietly emerged—albeit in very different form. Power for traditional media firms lies largely in defining and producing content for broad dissemination and consumption, but internet firms in large part do not participate in content production. (Or at least it is not a focal point for the central business model of consumer internet firms.) Google's value proposition is instead focused on offering the efficient and effective classification and location (searchability) of content, including news; Facebook's value hinges on seamless connection and engagement across a user's friend graph; and Twitter's is the attribution of ideas and engagement against those ideas by the broader user population. Across the industry the value lies in offering informed social connectivity, or "meaningful social interaction," as Zuckerberg has put it.

But it is not only provision of these services that distinguishes and strategically separates Google and Facebook from their competitors. If that were the case, there would be far more effective competition among and against these firms. A key part of their ongoing commercial strength lies in their "first-mover advantage" in seizing the reins of the consumer internet business model, premised on the creation of advertising exchanges, at the same time as two technologies—data storage and computing—were rapidly expanding.[66] Just as Google and Facebook settled on their advertising-based business models, these two technologies surpassed a key threshold that triggered the rise of the big-data economy, and as long as the internet companies could cling to its coattails, they would achieve great profitability.

It is the combination of these phenomena—the novelty of the targeted-advertising regime promoted across the media ecosystem alongside the coinciding rise in the capacity of the technologies enabling the rise of big data—along with their nominally unique consumer services, that affirmed their capitalistic syzygy and set leading consumer internet firms like Facebook and Google on their historic trajectory. What has emerged, though, is an economic logic that underlies the entire open web and that is algorithmically trained solely for the maximization of profit for the leading internet firms, subject only to a few blunt constraints.

Advanced machine-learning systems are implemented for gains in profit across the three pillars of the consumer internet business model, and in each of these core practices there is tremendous capacity for discriminatory results to be thrust onto the individual user. On a near-continuous basis, algorithms are trained to understand the consumer's preferences, beliefs, and interests, all of which are shuffled into the consumer's behavioral profile; to keep the user engaged on the platform by understanding and ranking all content existing in the realm of posts that could be populated in the user's social feed; and to push digital ads that are likely to engage the typical user. In a sense, then, learning algorithms are continuously and ubiquitously used by internet firms to infer as accurately as possible the individual user's true nature and, accordingly, what arrangement of content should be pushed at the user to maximize profits. It is a fluid operation—representing perhaps the most well-oiled machine the world has ever seen.

This is what I call the radical "commercialization of decisions," radical because of its continual refinement and complete ubiquity across leading internet firms. All decisions made by learning algorithms in the context of the consumer internet are now necessarily commercialized in light of the combination of supercharged big-data technologies and platform power. That is to say, each decision made by a consumer internet learning algorithm—be it determining what content to push at the user, making an inference about the user's character, or some other narrower practice—is financially incentivized and therefore initiated and propelled purely by the pursuit of commercial gains. There is currency tied to every decisionmaking process that occurs in the industry, no matter how impactful or important it is.

The consumer internet's vertical integration of boundless data collection and seamless logical integration of machine learning into every decision surface pertaining to every user and group of users is not discretized in the traditional hierarchical sense. It rather "ceaselessly establishes connections between semiotic chains, organizations of power, and circumstances relative to the arts, sciences, and social struggles," where there "are no points or positions . . . such as those found in a structure, tree, or root." This is the rhizome at work, as articulated by Gilles Deleuze and Félix Guattari in their *Capitalisme et Schizophrénie* writings. As with a rhizome, the ubiquitous commercialization of decisionmaking "has no beginning or end; it is always in the middle, between things, interbeing, intermezzo." It is an "alliance, uniquely alliance," and comprises "an antigenealogy. It is a short-term memory, or antimemory," in that no matter what data or currency may flow into or out from it, it has one governing, gravitational rule: profiteering. As in the rhizome, such boundless commercialization operates like water, in the sense that it "fill[s] or occup[ies] all of [its] dimensions" and is "defined by the outside: by the abstract line, the line of flight or deterritorialization," in that it expands its reaches to all commercial opportunities that exist under its regime of present technological circumstance and regulation. As such it exhibits the virtues of "expansion, conquest, capture, offshoots." As a byproduct, the consumer internet's force exposes "arborescent pseudomultiplicities for what they are," as the internet has done in the case of traditional media, governments, and other hierarchical institutions that have had their business models turned upside down. The consumer internet's nature, on the other hand, "has neither subject nor object, only determinations, magnitudes, and dimensions that cannot increase in number without the multiplicity changing in nature."[67] It is the ultimate exhibition of the actor-network theory suggested by Bruno Latour, Michel Callon, and John Law, a new technological paradigm exhibiting an endogenous *anima mundi* that connects users' personhood to the industrial server farm.[68]

The consumer internet's rhizomatic culture is one of discriminatory decisionmaking everywhere, in every circumstance, all the time—driven by the pure capitalistic energy of the Silicon Valley internet firm backed by the venture capitalists anticipating a barnburner buyout or initial public offering and the Wall Street executives hoping to cash in on any possible deal. This is a stunning distinction from prior times: the commercialization of decisionmaking has created novel opportunities to disseminate all kinds of speech—whether organic, commercial, or otherwise nefarious in nature—and inject it throughout the modern media universe. It has holistically diseased our long-standing traditions of independence and fairness. We have graduated from the formative "public good" conceptualization of the World Wide Web at its inception. We are now in the age of the "commercial good"—that is, the good of the firms leading the industry. That such radically granular profiteering has become the *lingua franca* of American commerce, the new Weltanshauung of the business world's cutting edge, is quite a novel social circumstance. The media ecosystem of the twentieth century, in contrast, did not involve the commercialization of the multidirectional churn of fine-grained information. This was perhaps true even in the early stages of the internet through the turn of the millennium. But now algorithmic developments, including the deployment of sophisticated learning models by the most cash-rich firms in the world—alongside their data-gathering practices and advantageous pseudomonopolistic positions in a market with a paucity of true or would-be competitors—have collectively introduced a vicious circumstance by which companies such as Facebook have the opportunity to initiate, advertise, and host a market for information dissemination in such a way that those willing to pay to play will gain the opportunity to unite with this diabolical rhizome that manipulates the individual's experience of the world. They have treated the human psychological condition as commercial terra incognita.

That is not to say that this centralized decisionmaking power did not characterize past instantiations of the American media ecosystem as well. The prior world dominated through broadcasting, radio, and print news institutions, too, and had the capacity to produce and perpetuate bias. But there were key differences. Their mined data were not as granular or personalized, owing to the nature of the technology in question; a consumer internet laden with learning algorithms evolving and operating over corporate servers and producing results within milliseconds on Google's search page generates entirely different impacts on the citizen. Furthermore, the more traditional past instances of the media ecosystem were heavily regulated, either directly, by the government, or indirectly, through a combination of measures instituting industry-wide transparency and public accountability.[69] Examples include federal election regulations for the broadcast and radio formats as well as standards syndicated across the journalistic community, including the Society of Professional Journalists' Code of Ethics, the preamble of which proclaims that "public enlightenment is the forerunner of justice and the foundation of democracy. Ethical journalism strives to ensure the free exchange of information that is accurate, fair and thorough. An ethical journalist acts with integrity."[70] And should a journalist not, he or she will be ousted from the industry by consensus.

Thus, in practice, the capacity for traditional media firms of the past to engage in unfair practices leading to potential consumer harms was constantly policed. While they did nevertheless have tremendous power—as these formats collectively constituted the media ecosystem—they experienced continual pressure and possessed limited capacity to syndicate damaging impacts on the public.

Individual capacity to determine what we see and are subjected to has been holistically undermined and diminished by the consumer internet firms. Whereas the venue for information in decades past featured an open space for independent thought, it has now been invaded by a silent form of commercial speech; the content displayed before us comes at the beck and call of the firm responsible for populating the results page. Each time we open the laptop or check the phone and use the services central to information consumption, we are subjected to an array of information preselected and ordered for us at the determination of a mercenary machine, with nothing trained into its decision modeling except profit maximization. And as researchers Gordon Pennycook, Tyrone Cannon, and David Rand have found, "even a single exposure [to fake-news headlines] increases subsequent perceptions of accuracy, both within the same session and after a week."[71] Such virulent consumption of presupposed conclusions within the filter seems to go hand in hand with related findings by Soroush Vosoughi, Deb Roy, and Sinan Aral that, based on analysis of rumor cascades from 2006 to 2017, fake news travels faster and further than the truth on Twitter.[72]

Constitutional legal scholar Cass Sunstein contends that human minds were not meant to deal with this kind of capitalistic ease. We are biologically trained, he contests, to see a wide unlabeled array of content and contend with its merits and demerits to the end of deciding for ourselves whether we shall take up and believe the objective information or opinion-driven arguments contained therein.[73] This is Henry David Thoreau's civil disobedience—the Twitter feed has subverted the very idea of independence of the mind and oppressed humanity's interest to such an extent that it is not only our democratic process and progress but also our intellectual future itself that faces direct and immediate threat.

To examine this conundrum from a different angle, it is the third layer—the content dissemination network—of the internet-media industry that has now been radically industrialized. The other two layers—the physical network infrastructure and the content—were already industrialized in decades past. It could be said that the third layer of the infrastructure should never have been a free market in the first place, at least not as it is now. Leaving aside whether and how much the first two layers should have been opened to the industry and focusing only on the third, we can observe that the industrialization of dissemination apparently subverts the consumer's interest if left to the free market, given such perversions of the media space, including the dissemination of disinformation and the wide spread of hate speech. The third layer's industrialization has a more direct influence on our thinking than the others; it is a novel form of mind control.

Professor Helen Nissenbaum argues that the approach to consumer privacy protection undertaken by the Federal Trade Commission and Department of Commerce is dangerous, noting that the U.S. government's "interest has been limited . . . by a focus on protecting privacy online as, predominantly, a matter of protecting consumers online and protecting commercial information; that is, protecting personal information in commercial online transactions. Neither agency has explicitly acknowledged the vast landscape of activity lying outside the commercial domain."[74] Nissenbaum is referring to the manner in which U.S. governmental agencies focus not on privacy concerns at large as they occur across society, including within governmental agencies and regulated entities such as hospitals and banks, but rather only on those occasions when the data transfer affects "consumers"—those individuals party to some identifiable monetary transaction in the marketplace. Based on the foregoing discussion, we can extend Nissenbaum's point to the lack of effective oversight over the commercialization of decisionmaking—precisely because the narrow and independently minor decisions made using the classifier models developed by learning algorithms typically do not have dollars directly attached to them. No one is paying into Facebook's rhizomatic regime for clearly defined and discretized forms of prioritization. But these systems are nonetheless designed to yield the greatest possible profit margin for the service operator—and even perpetuate provably discriminatory decisions against individuals and classes of individuals so long as doing so remains nontransparent to the public and is aligned with the firm's profit motive.

To that end, the collection of personal information is ever present, and its transfer among firms involved in the digital media ecosystem is multidirectional. Indeed, the modus operandi of leading internet firms is at once to be at the center of and to reach its tentacles across the information-sharing network that sprawls the digital ecosystem. Firms like Google accordingly use a multitude of technologies and technological protocols to collect personal data, over their own platforms as well as through web cookies and physical equipment technologies.[75] Critically, this information is maintained by the firms within private, walled gardens—proprietary systems that maintain their collective hegemony over the knowledge of customers' individual profiles for content-targeting purposes. Furthermore, firms "lease" the information out in anonymized formats, enabling advertisers to target certain classes of the population at will. Sometimes, the advertiser might inject its own data into Google's advertising platform, encouraging Google to help it reach audience segments to a remarkable degree of precision. This bidirectional relationship is critical to the functionality of the consumer internet—and operates as the grease at the joints of the industry, enabling the radical commercialization of decisionmaking.

The Ubiquity of Bias in Consumer Internet Platforms

The commercialization of decisionmaking in the consumer internet has been shown to be detrimental to marginalized groups, including protected classes of the population. When markets elevate currency over values as they do, the resulting economic logic enables the pursuit of the highest possible profit margin at the expense of any other concern, particularly if it is unpoliced and extends over a largely unregulated market such as the internet. Machine learning is the tool that enables the collation and exploitation of information, thus reducing transaction costs even further, and the profits generated thereby typically are drawn up by the industrial entities responsible for implementing the learning models in an integrated manner. Indeed, the internet is highly effective as a means for communication because it reduces transaction costs in the exchange of information relative to the traditional media of the past, which did not enable personalization of rendered services or collection of information on the consumer in the first place. The internet thus enables a two-sided exchange in a manner we had no capacity for in years past.[76]

But it is precisely this reduction of transaction costs that has enabled discriminatory outcomes that disfavor marginalized communities, particularly in the United States, where the internet is in such wide use, the internet industry has maintained tremendous political power, and the demographic heterogeneity and political economic tradition and trajectory are such that the capacity for internet-enabled discrimination has become supercharged.

Targeted Advertising Regimes

The institution of the commercial regime underpinning the consumer internet economy—targeted advertising—has enabled both intentional and unintentional discriminatory outcomes. Typically, ad-targeting regimes take advantage of the commercial interests of two types of parties: the advertisers that wish to communicate their products and services to consumers and persuade purchasing decisions and the platforms and publishers that have access to consumer attention and therefore own ad space.[77]

Usually, platforms also possess and analyze large, refined stores of information on consumers. The raw data might include data collected about the consumer's on-platform activity, including what products, social media posts, and search results the consumer interacts with; the consumer's off-platform web activity pertaining to activity on third-party websites, including mouse clicks, browsing pathways, and content consumed; location information shared with the platform via the consumer's smartphone, should the consumer have opted into location sharing with the platform service (or through other means in certain cases);[78] location and behavioral data collected through other device technologies, such as beacons and routers that interact with the consumer's devices in the physical world; and data purchased from or voluntarily shared by third parties, such as data brokers and advertisers.

Advertising platforms—including those implemented and hosted by Facebook, Google, and Twitter—take advantage of such data-collection regimes to infer behavioral advertising profiles on each user using the company's internet-based services. Those behavioral profiles are maintained by the platform firms and remain largely nontransparent to third parties. But should advertisers such as apparel designers and retail banks wish to target certain audience segments—for example, young people in Italy, Portugal, and Spain who are fans of Cristiano Ronaldo and have a certain income—the platform firm typically analyzes its data stores and determines which grouping of consumers in the target geography would be likeliest to purchase the advertiser's wares. Clearly it is this determination of who should go into the targeted audience segment that engenders harmful disparate impact. A recent suit by the U.S. Department of Housing and Urban Development illustrates this tension clearly. The department used its authority under the Fair Housing Act to allege that Facebook enables harmful disparate impact in making available housing opportunities because advertisers can target certain groups according to their membership in various consumer classes, including protected classes such as race and gender.[79]

What is perhaps most dangerous is that civil rights laws in the United States designed to ensure a modicum of economic fairness in housing and employment cover only certain critical industry areas. Unfortunately, such protections against a commercial operator enabling disparate impact in the majority of other areas does not necessarily trigger a civil rights violation, despite the clear discriminatory outcomes that can arise from shady scams targeted at certain marginalized communities or, conversely, favorable ads such as investment opportunities being pushed exclusively at more mainstream communities.

Meaningful Social Interaction

Consumer internet firms deal in a novel form of currency: the collective combination of the user population's personal information and attention. By raking as much of this as possible and amassing it to generate collated ad space that can be sold off to the highest bidder via intelligent auctions for the purpose of enabling targeted commercial speech, the internet companies maximize their value proposition to businesses that wish to advertise back at the consumer. It is a vicious cycle that takes advantage of many market desires—in particular, the need to continually engage users so that they spend as much time on the platform as possible and engage with it to the greatest possible extent.

In 2018, Mark Zuckerberg proclaimed that his company would institute new changes to the algorithm driving the social media network's core news feed service that ranks the universe of content available to a given user on the home screen. He noted that the company would now focus on promoting "meaningful social interactions."[80] That is not to say that this was not always in the company's designs: he discussed how recent events had illustrated more clearly that there was too much passive interaction with content, particularly posts shared by "businesses, brands, and media."[81]

But what does meaningful social interaction really entail? Conveniently for Facebook, it is a metric that, if effectively maximized, can contribute to the two resources it principally cares about: the user's attention and his or her personal information. Effective meaningful social interaction would keep users on the platform by connecting them to more personal social content that they actually want to see—and if they engage more with such content, Facebook will know it and thereby know their users better, such that ads can be disseminated more efficiently at them and ad space can be increased.

This is where the power of commercial machine learning—and resulting machine bias—come in. There is no foolproof way to determine what types of content matter to an individual user. Broad inferences can be drawn, but it might be challenging to infer what academic subjects and scholars resonate with an individual, or which particular players on an NBA team or which shade of blue a user likes the most. This is the fallacy of data and, by extension, learning models: they are used to estimate the real feature but cannot ever offer a precise representation of the real world—and yet they are continually used to make determinations about what an individual actually cares about in the real world. Thus the leading consumer internet companies' efforts to enable meaningful interaction—whether in the context of a search engine or e-commerce platform or social media network—is flawed at best. Next time you are on Instagram, check out your Ad Interests—the conclusions Facebook has drawn about your personal interests (and which it uses to target ads at you) based on your Instagram engagement.[82] It is likely that you will see interests that don't really interest you.

The industry's use of highly sophisticated artificial-intelligence systems, including neural networks for real-time analysis of user behaviors—in conjunction with social science research conducted within the industry itself—powers the refinement of the models used to rank such features as a user's Instagram feed. But regrettably, such systems have the propensity to supercharge the deployment of assessments about the user in ways that misrepresent his or her interest. If a user does not interact with some mundane piece of content because it does not personally resonate at a social or intellectual level, the platform must reorient its assessments about that user. It is this dynamic that has led Facebook to group individuals by political allegiances[83] and has kept YouTube from screening inappropriate videos.[84]

These platforms constitute an online commercial playscape in which disparate impacts can run riot—where only certain marginalized classes are shown (or not shown) certain forms of content. Even if it does not constitute a civil rights violation, the content dissemination to a given user class might damage the economic prospects of an individual in the group. If Facebook decides that an individual is more likely to be interested in new Netflix shows than in the study of microeconomics, for example, that person might never be offered content that would encourage him or her to adopt better practices around personal finance, better awareness of the political state of the nation, and better awareness of broader economic opportunities that might be available, should the person know where to look for them.

Whether such ranking models are fair ultimately depends on the design of the algorithm that maximizes so-called meaningful social interaction, defined by and algorithmically trained to service the commercial objectives of the platform operator.

The Pursuit of High-Value Audiences

There is a tradition in Silicon Valley, particularly in the consumer internet industry, whereby fledgling firms tend to serve those niches that are already financially secure. Because their efficacy is demonstrated by serving those high-value customer segments, firms might receive investment funds to tackle broader growth as well. Indeed, companies leading the sector have variously been party to such practices: Facebook first invited only Harvard students to participate on the network;[85] Airbnb initially served only those cities where real-time hotel prices were high;[86] and Gmail's beta version was distributed first to a select group of people and those friends they wished to invite to use the service.[87]

Needless to say, such communities—elite universities, would-be hotel patrons in rich cities, public intellectuals, and the like—are not representative of any community beyond the intelligentsia and tend to deprioritize the needs of marginalized communities that are most often subject to harmful discrimination. Nonetheless, it is through the observation of these initial groups' interactions with the platforms that engineers attempt to design the form their platforms will take in an eventual steady state. This culture of serving the privileged first before rollout to the rest of society is seemingly part and parcel of the investment culture that breeds the most well-known American venture capitalists.

It is a culture that considers lower socioeconomic classes last. And when overlaying the development and refinement of learning models over this dilemma, the potential for machine bias to lead to disparate impact becomes resoundingly clear. The argument could be made that the platform firms safeguard against this potential harm in various ways—for instance, by protecting against built-in propensities for learning models to perpetuate biased outcomes—but the design of the platforms necessarily must favor the elite and wealthy. In a capitalist regime favoring free markets, no other approach is viable for venture capitalists and founders; if they do not take advantage of the economic opportunity of serving the well-off customers first, the competition will eventually do so and overtake them. If at any point Facebook were to lose high-value users to the competition, the company would have to either acquire those competitors or reorient the ways in which the platform fundamentally works so as to increase the probability that high-value users might come to the platform. This is exactly the strategic circumstance Facebook finds itself in now as it considers how to claim or reclaim the the young users who opt for Snapchat, Instagram, and YouTube over Facebook.[88]

Considerations toward an Ethical Approach to Designing Algorithms for the Future

The business model that sits behind the front end of the internet industry is one that focuses on the unchecked collection of personal information, the continual creation and refinement of behavioral profiles on the individual user, and the development of algorithms that curate content. These actions all perpetuate the new pareto-optimal reality of the commercial logic underlying the modern digitalized media ecosystem: every act executed by a firm, whether a transfer of data or an injection of content, is by its nature performed in the commercial interests of the firm because technological progress has enabled such granular profiteering. The firm is a rhizome; it faces no obstacles to force biased outcomes on users because there is typically no legal or regulatory repercussion for formulating that bias. This novelty in the media markets has created a tension in the face of the public motive for nondiscriminatory policies; where adequate transparency, public accountability, or regulatory engagement against industry practices are lacking, it is directly in the firm's interest to discriminate, should discriminatory economic policies suit its profit-maximizing motive. As elucidated by professor Sandra Wachter, there are many gaps in the law, which is "ill-equipped" to address the problems associated with the kind of affinity profiling facilitated by typical digital advertising platforms.[89]

What distinguishes the consumer internet sector is that it is not subject to as rigorous a regulatory regime as the telecommunications, healthcare, and financial industries are. It is in a state of choiceless awareness; the operation of online digital services over a physical infrastructure is still largely a novel practice as far as the laws are concerned, and the U.S. Congress has not yet acted. In this mostly unregulated environment, these firms have had the opportunity to grow profits through a combination of business practices that most effectively yields highest margins—in just the way that Karl Marx suggests capitalists would.

Of particular concern is the currency these firms deal in—consumers' personal information and attention—and the opaque mechanisms by which they rake in profits. Some industry executives argue that the services they offer are free—a misleading conjecture. True, consumers do not pay monetary fees for their services, but as the next chapter, on the competition pillar of the business model, explains, the most effective consumer internet firms amalgamate a complex combination of user attention and data on the end-user side of the market and translate it through an automated digital advertising exchange into monetary reward in the advertising market.

FOUR

Platform Growth

Capitalism Consuming the World

The enforcement of U.S. public policy involving competition has dramatically regressed from a once aggressive stance to a largely noninterventionist state over the past several decades. That regression has coincided with an increase in market concentration of the principal firms in key industries across the American economy.[1] For no industry is this case in as stark resolution as for the consumer internet—in particular, for the three-leading digital platforms: Facebook, Google, and Amazon.

These three firms' unscrupulous and unprecedented climb to market domination has ignited fresh policy debates over the need to return to a competition-regulatory framework to encourage competition.[2] While most firms agree that their economic power affords them a level of control over consumers that must be curbed, there is vociferous debate as to what American policymakers should do.[3] Many, including presidential candidates, have famously called for breakups.[4] But as competition policy expert Phil Verveer has suggested, the practical considerations of pursuing hard-line antitrust policies abound.[5]

Those concerns about our regulatory system are addressed later in this chapter. But to leave aside the broader question of how we should apply a robust competition framework to modernize the method by which we regulate the leading internet firms would be folly. As will be discussed in this chapter, the business model of the consumer internet is tightly hinged on the creation of highly compelling platform services such as Instagram and Google Search—services that at once addict the consumer and shut out the possibility of rivalry. Indeed, this is one of the three pillars in our conceptual framework of the business model: the leading internet firms proactively seek ways to draw consumers in, riding the network effect to the point that they gain mass popularity. They then swoop in to kill off the competition in whatever way possible—competition that might otherwise also have been able to participate in the market for consumer attention.

The overt exhibition of anticompetitive practices by such firms as Facebook has been left untouched by the American regulatory process.[6] Unless this unacceptable situation is corrected, the individual consumer and citizen will in perpetuity suffer rank economic exploitation.

The New Contours of the Digital Advertising Industry

Over the past few years, some have challenged the notion that the lion's share of the digital advertising market will securely remain the dominion of Facebook and Google for the foreseeable future, as a new player has arrived on the scene: Amazon. Amazon's emergence in digital advertising represents a major challenge to the share of digital-ad spending that Facebook and Google have enjoyed for many years now.

The proof is in the numbers: whereas Amazon's market share totaled around just 2 to 3 percent in 2016, that number grew to 6.8 percent in 2018 and was projected to grow to 8.8 percent in 2019 and to surpass 10 percent in 2020.[7] This precipitous jump—which puts it solidly in third place behind Facebook and Google—does not appear to have been induced by an artificial blip or an accounting update; it is the result of a concerted strategy undertaken by Amazon to tackle the digital-advertising ecosystem, which, if done right, can generate high-margin profits into the future. Facebook and Google, meanwhile, enjoyed a massive hold on the digital-advertising market in 2017, although that share is projected to drop to less than 60 percent in 2020.

Amazon's emergence in digital advertising appears to have come quite late in its maturity as a company; it is a behemoth of a firm, with the lion's share of a major U.S. consumer market—e-commerce—and the platform visibility and audience that comes with that market dominance. Amazon's 2018 revenues amounted to more than $230 billion—an eye-popping 1.1 percent the size of the entire U.S. economy.[8] The vast majority of that bounty comes from its signature shopping service, a presence that trumps that of any other American e-commerce firm. Through that platform, Amazon has generated the capacity to collect inordinate amounts of highly valuable data on the Amazon user. This silent but steady churn of data collection is executed primarily through user engagement with Amazon's key services, including its shopping service, where customers can search for anything from Audemars Piguet chronographs to zipper-tab replacements; Amazon Prime, where customers can stream many of the latest media offerings, including television series and films; and Alexa and Kindle, its virtual assistant and e-reader devices, respectively. In addition, Amazon purchases personal information from data brokers, such as Equifax's TALX,[9] although it is difficult to know the extent to which this is happening, since Amazon does not reveal what quantities of data it buys and which data brokers it engages.

This combination of highly trafficked consumer platforms and ubiquitous data collection has encouraged Amazon to think more creatively about how to enter the digital-advertising space. It controls a tremendous consumer audience through its dominating e-commerce platform, and it has inferred the individual preferences and desires of every individual user of that platform. This compelling combination forges the path to enormous success in the consumer internet advertising market.

Why did Amazon not place its bet on digital advertising years earlier? There are most likely a combination of reasons: that Amazon was more intently focused on other areas, including the expansion of its e-commerce platform, which by today is a settled cash cow, or the buildout of its Amazon Web Services commercial cloud-computing offering, which contributes just 11 percent of the firm's overall revenues but around half of its operating margin;[10] that it wanted to give its platform service more time, perhaps after one of its recent marketing pushes, to settle into broad use with the mainstream public; that it was fearful that introducing display advertising alongside its core services, especially for Prime members, could jeopardize its relationships with its customers; that it saw advertising as a more compelling opportunity only after witnessing the policy troubles that Facebook, Google, and Twitter have encountered since late 2016; or that it simply did not possess the ingenuity or expertise needed to construct a robust digital-advertising platform until recently.

But Amazon is fully invested in advertising today—and it is here to stay. Some other platform firms have also made noteworthy strategic moves in the digital advertising market. Snap's share, for instance, has also grown precipitously, with overall advertising revenues rising from $600 million to $1.2 billion in 2018.[11] Twitter,[12] Verizon,[13] and AT&T[14] have each also made recent moves in the industry with varying success. Still, the efforts beyond those of Amazon remain relatively minor.

From Duopoly to Triopoly amid the Emergence of a Novel Digital Currency

Many conclude that Amazon's surge in the digital-advertising market constitutes real competition to Facebook and Google. For those two firms, watching Amazon's steep climb to 10 percent of the market in such a short time will doubtless have raised eyebrows among their own advertising executives. But such fears should be discarded out of hand.

Consider the possibility that Amazon's presence in the market will not challenge Facebook's or Google's practices, nor will Amazon's emerging digital-advertising practice itself be challenged by Facebook or Google. We have already seen the evidence of this in the period that Amazon's market share has climbed; the three firms appear to be completely settled in their newfound positions within the fast-expanding market of digital advertising, and none of them needs to make strategic calls to beat out the others in attracting advertiser interest. It is accordingly difficult to imagine that Amazon's emergence will induce Facebook and Google to vastly change their own practices—in the ways that they curate content, target ads, or collect data on individual users—in the manner that we as consumers might desire.

We are witnessing the quiet emergence of a new three-way hegemony over digital advertising in the United States led by these firms. Why? I contend it is because online advertising is a wholly different sort of market. Yes, it has cash-rich patrons, the most revenue contributing of them being Fortune 500 businesses, including Proctor and Gamble, Toyota, Samsung, and Apple.[15] But the currency that the digital platforms receive from advertisers—indeed, their direct monetary revenue—is a mere fraction of the actual payment they receive for their core services, whether they are social media platforms or search engines or e-commerce platforms. The actual payment they receive is delivered by their end consumers—the users of Instagram and Google search. And that payment is delivered in an entirely novel form of currency that has only emerged recently with the rise of the big-data economy: the complex combination of our attention and our personal data.

Here is the rub: the advertising revenue contributed by marketers is simply the manner in which the digital platforms exchange the combination of our attention and data for monetary value. It is our attention and data that these three firms care about more than anything else; they know, given their dominion over the web, that once they have captured the consumer audience and an accurate set of detailed behavioral inferences drawn against every individual in that audience, they can easily monetize that powerful combination in the advertising market.

Let us consider that, with respect to their advertising businesses, Amazon, Facebook, and Google are two-sided platforms.[16] On one side of the platform, they deal with us, the end consumers. On the other side, they deal with marketers who want to place targeted ads to convince end consumers to buy something or take some action. On the consumer side of their platforms, each of the three firms enjoys a market monopoly—Google in search, Facebook in social media, and Amazon in e-commerce. As such, each firm can charge monopoly rents in this novel currency that combines our attention and data. On the marketer side of their platforms, Amazon, Facebook, and Google transfer the attention and data that were collected at monopoly rates from the other side of the platform and exchange it for cash. Because the three firms enjoy monopolies in their respective industry silos, each can charge monopoly rents that are raked in the form of our attention and data—which explains the historic profit margins that their advertising businesses yield. Note that this novel currency should not be viewed as a nonrivalrous good. Without a doubt certain features of the consumer internet industry are antirivalrous, as U.C. Berkeley professor Steven Weber has put it, whereby as more people share the antirival good, the greater value each party appreciates. This is the case for certain types of software—like the Instagram app, for example—and for many forms of data. Another example exists in the case of spoken language. As Lawrence Lessig of Harvard University notes: "not only does your speaking English not restrict me, your speaking it benefits me. The more people who speak a language, the more useful that language is, at least to those who speak it. We therefore don't tax foreigners who learn our language; we encourage them, since we benefit from their use of it just as they do." But this novel currency that combines our data and attention is economically intelligible only to the service provider, so that there is a limited supply of it. The novel currency is not antirival; we can think of it as a novel extension of our personhood.[17]

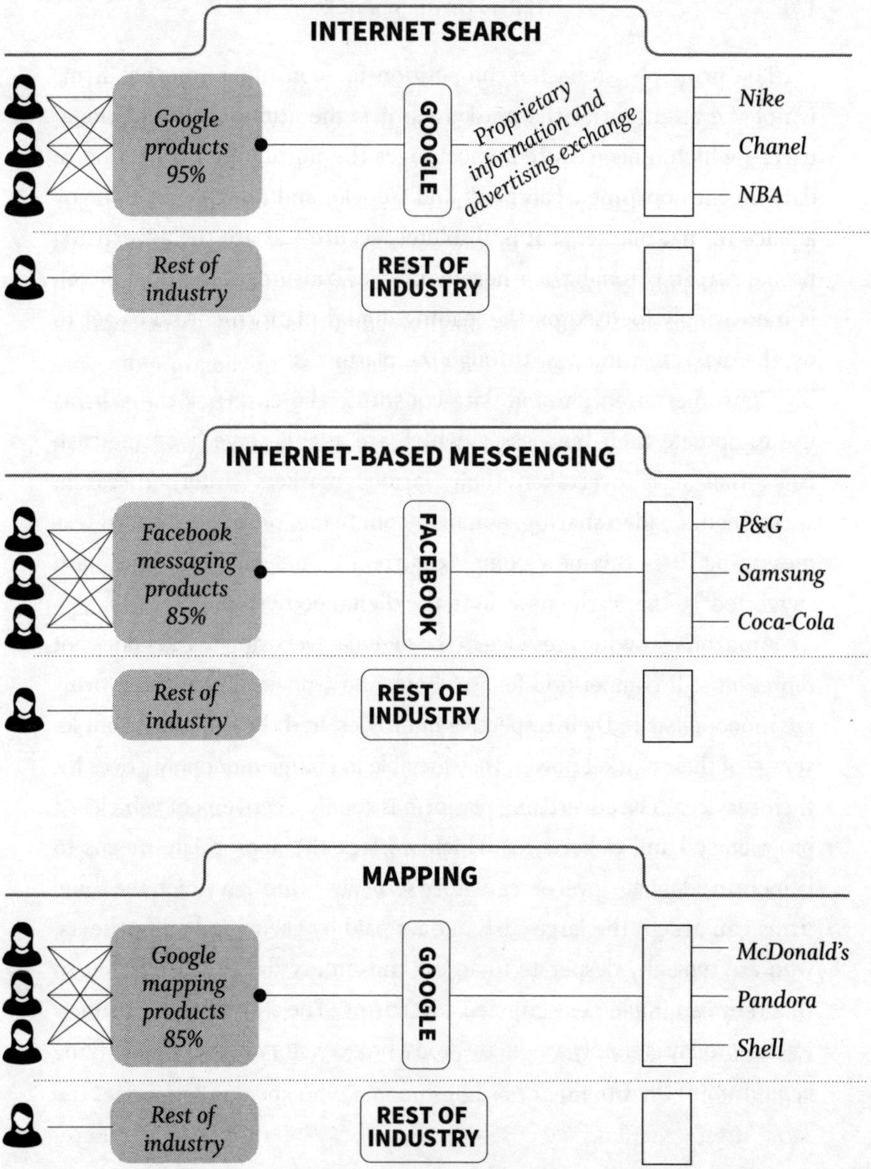

The monopoly position of the leading internet firms, which are two-sided platforms, enables the extraction of rents from consumers as a novel currency of data and attention that is only economically intelligible to the digital platform that coordinates its in-context collection and monetization. With this monopoly rent in hand, the digital platform transfers the underlying economic value to the other side of the platform through an advertising exchange and earns a high margin in profits directly from marketing clients, such as consumer brands.

The principle arena for competition for consumer internet firms is not the attention of the marketers; it is the attention of individual users—which, once secured, encourages the uninhibited collection of data on the consumer. Facebook and Google, and now Amazon, enjoy a place in that market that is absolutely secure—as are the advertising profits reaped through their hegemony over consumer attention, which is increasingly focused on the leading digital platforms. And to get to us, the marketers must go through the platforms.

Consumer attention and data constitute the currency these firms use to operate their businesses, which are at this stage booming cash cows that have overtaken their various markets—internet search, social media, video sharing, e-mail, e-commerce, or internet-based text messaging. It is this new complex currency—not data, as some have suggested[18]—that is the new oil in the digital ecosystem.

Amazon's growing presence in the digital-advertising market does not represent stiff competition for Facebook and Google. These three firms are monopolists in their respective industries, and through persistent leverage of their market power, they are able to charge monopoly prices for their services. The advertising platform is simply a convenient vehicle—a proprietary bank of sorts for which we lack the appropriate means to launch an adequate level of regulatory scrutiny—through which the three firms can accept the large cash outlays paid by their marketing clients, who are typically desperate to access consumers' data and attention so that they can implement targeted advertising. The story of the consumer internet today is simply a vignette atop a broader story of ever-intensifying capitalism in the United States.

The Exploitative Rake of Data and Attention

The collection of our attention and data is routinely executed by Facebook, Amazon, and Google, and is done so painlessly for the internet firm and for the end consumer, unlike payments made in the physical world, where we at least need to take out our wallets. As scholars

Top Ten Social Media Platforms among U.S. Internet Users, Ranked by Market Share of Visits, August 2019

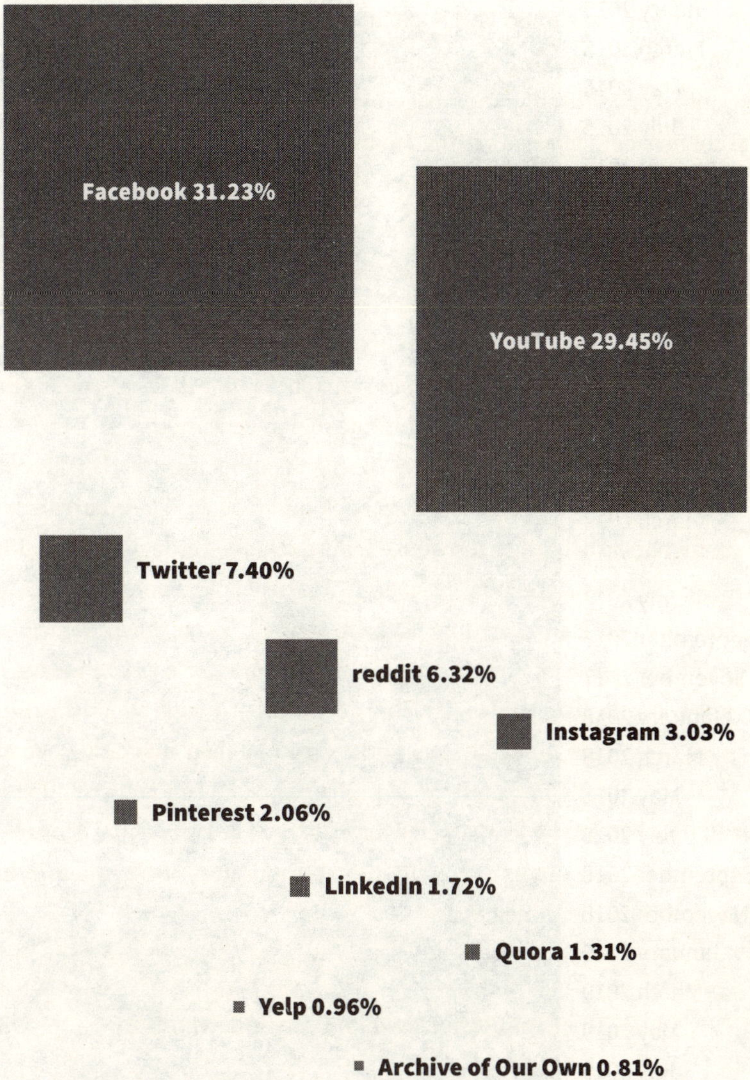

Note: Desktop and mobile.

Source: Hitwise, September 3, 2019 (www.emarketer.com/chart/230531/top-10-social-media-platforms-among-us-internet-users-ranked-by-market-share-of-visits-aug-2019).

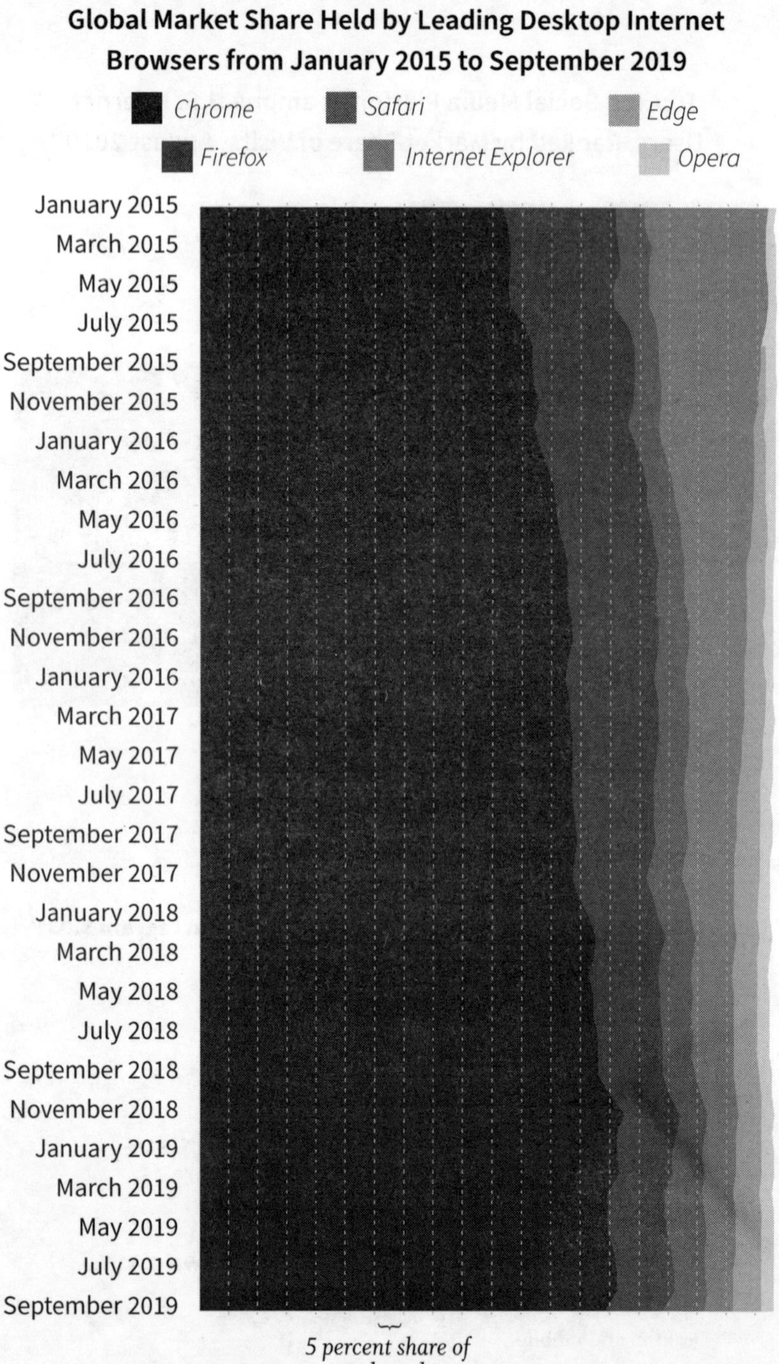

Source: Shanhong Liu, Statista.com, October 15, 2019 (www.statista.com/statistics/544400/market-share-of-internet-browsers-desktop/).

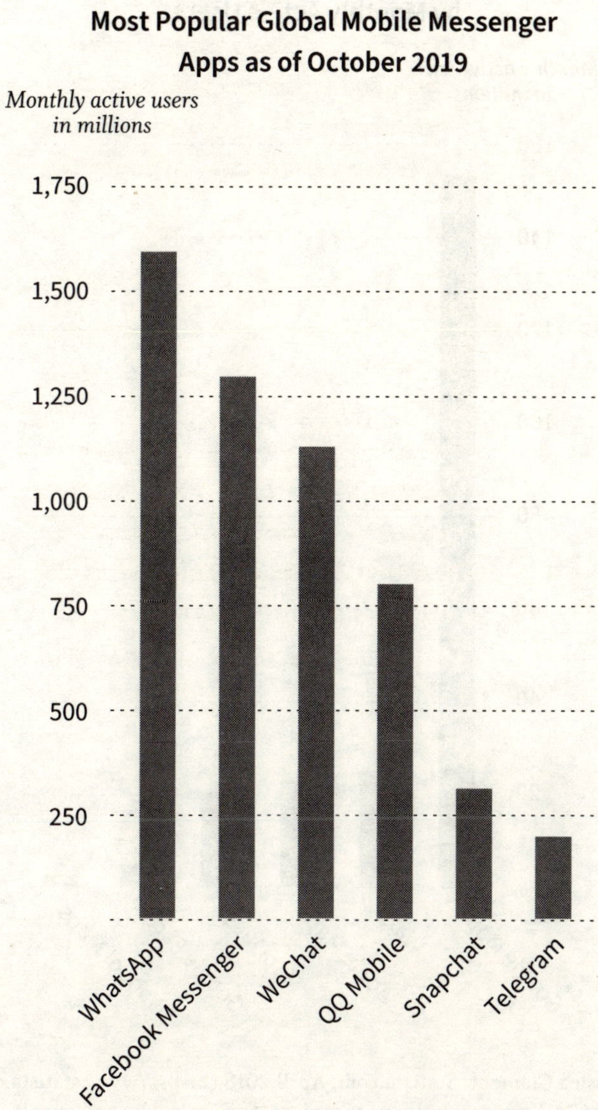

Source: Jessica Clement, Statista.com, November 20, 2019 (www.statista.com/statistics/258749/most-popular-global-mobile-messenger-apps/).

Market Shares of Mapping Applications by Monthly Active Users

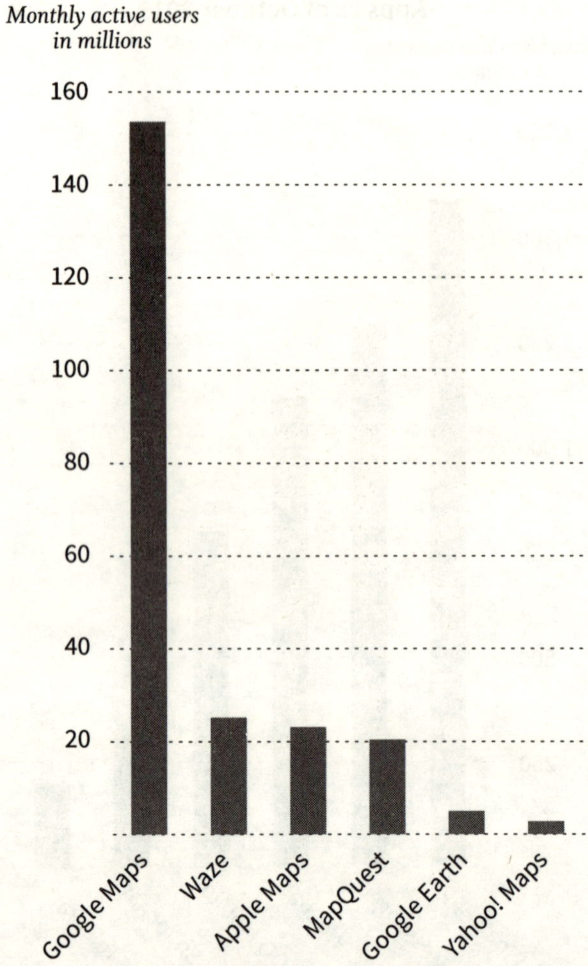

Source: Jessica Clement, Statista.com, April 2018 (https://www.statista.com/statistics/865413/most-popular-us-mapping-apps-ranked-by-audience/).

Market Shares of Total Over-the-Top Online Video Viewing

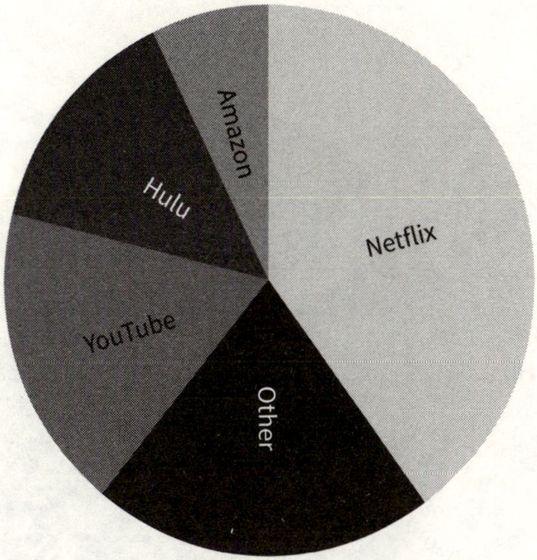

Source: Ayaz Nanji, "The State of OTT Video Viewing: Top Devices and Platforms," MarketingProfs.com, June 2017 (www.comscore.com/Insights/Presentations-and-Whitepapers/2017/State-of-OTT).

Global Platform Market Shares in Demand for Original Digital Content

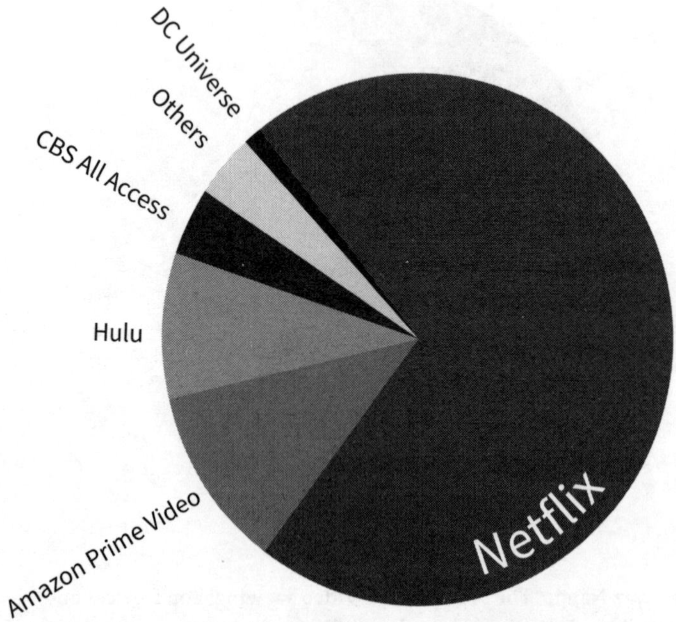

Source: "Global SVOD Market Share Trends Based on Audience Demand for Digital Originals," Parrot Analytics, May 2019 (www.parrotanalytics.com/insights/global-svod-market-share/).

Top 10 U.S. Companies Ranked by Retail Online Sales Share in 2019

Rank	Company	% Retail Online Sales Share
1	Amazon	47.0
2	eBay	6.1
3	Walmart	4.6
4	Apple	3.8
5	The Home Depot	1.7
6	Costco	1.3
7	Wayfair	1.3
8	Quarate Retail Group	1.3
9	Best Buy	1.3
10	Macy's	1.2

Source: Marianne Wilson, "Amazon to Capture 47% of All U.S. Online Sales in 2019," Chain Store Age, February 2019 (https://chainstoreage.com/technology/emarketer-amazon-to-capture-47-of-all-u-s-online-sales-in-2019).

have noted, consumers do not immediately ascribe a fair market value to their personal information in the moment of its collection. The consumer lacks sensitivity to the collection of personal information—or, as my colleague Ben Scott and I have noted in another work, consumers lack understanding of the "privacy prices" they are subject to.[19]

As a result, these "privacy prices" are perfectly inelastic—meaning that to get access to a consumer internet service such as Google Search, we do not care how much of the currency in question (how much of our personal data and our attention) we will have to shell out. This dynamic persists for two primary reasons: average consumers lack knowledge and understanding of the fair value of their personal information,[20] and their experience in submitting to such prices, extortionate as they may be, is frictionless.[21] Customers do not experience the collection of their data or the sale of their attention; they have simply downloaded all of their favorite apps, apparently for free. The rake of customers' currency goes unnoticed as they swipe right and left, scroll through the potential friends Facebook has suggested for them, and conduct one news search after another to read the latest reports about Charles Leclerc's contract situation with Scuderia Ferrari. As long as the aggregated collection of our data and attention remains frictionless to us, whether because of our ignorance or the industry's ingenuity, the rational action for firms that operate digital-advertising platforms is to collect as much of it as possible. This is particularly the case because we are land-poor in the digital landscape; we possess the resource of our data and attention but are incapable of extracting a fair amount of economic value from it because we lack access to the interpretative technological infrastructure possessed by the consumer internet firms that can analyze the collection of our data and attention on an automated basis and exchange it at monopoly rates for advertising revenue.

Companies like Facebook and Google have nothing to fear from Amazon's increasing market. The industry overall is expanding rapidly, including the absolute advertising revenues that Facebook and Google appreciate each year. Despite the decline in their aggregate share in the digital advertising market, both Facebook and Google have experienced double-digit percentage growth in advertising revenues year over year in the interim, and the overall market in digital advertising has correspondingly grown at a fast clip as well. Amazon's market growth does not pose a threat to Facebook or Google because of the firms' absolute control over the markets they are most invested in—the markets for consumer data and attention. Facebook and Google are much less concerned about the direct advertising proceeds from that data and attention—they know it will come because they know that wherever public attention lies, the marketer will have to chase consumers. Meanwhile, their presence in the markets for internet services—social media and internet search—is secure, as is the magnitude and value of the currency they extract from end consumers. The firms' domination of the digital data and attention markets and the resources they bear represent, for the user, a *fait accompli*.

Seen in this light, the industry's argument constitutes an elaborate attempt at erecting a Potemkin village to throw the policymakers and public for a loop. Amazon is participating in a new three-pronged dominance of the digital-advertising market, largely determined by the reach and effectiveness of the internet services that these companies develop. This suggests that Amazon's growing presence in the digital-advertising market is nominal at best. It does not threaten either Facebook or Google; on the contrary, it does little to shift the competitive dynamics in the digital-advertising industry at all.

The Fallacy of "Free" Value over the Consumer Internet

Industry executives often say that their services are free. This is completely misleading.[22] The currency paid by consumers for Facebook's service comes not in the form of money but rather from the complex combination of data and attention. And while Sheryl Sandberg has taken the opportunity to apologetically clarify that Facebook "depends on your data," it took the Cambridge Analytica incident to force her to publically do so on NBC.[23] Industry executives avoid this sort of public attention; when called to task over their companies' questionable business practices, they have to reveal the nature of their business models—and risk exposure of the evils those business models engender. Such admissions inject increasing amounts of friction into the consumer internet firms' presently frictionless rake of our attention and data.

The result is that Facebook, Google, and Amazon force an implicit business practice down the user's throat—the boundless collection of personal data. No federal law in the United States restricts their capacity to do so. Sandberg, in the same NBC interview, touched on this, noting that "we don't have an opt-out at the highest level. That would be a paid product."[24]

Nonetheless, Facebook's take-it-or-leave-it proposition has been implicated in a recent charge by the Federal Cartel Office of Germany, which brought a novel case against the company in 2019. The Cartel Office—or Bundeskartellamt—argued that Facebook has a monopoly

Market Shares in Digital Advertising (2017–2018)

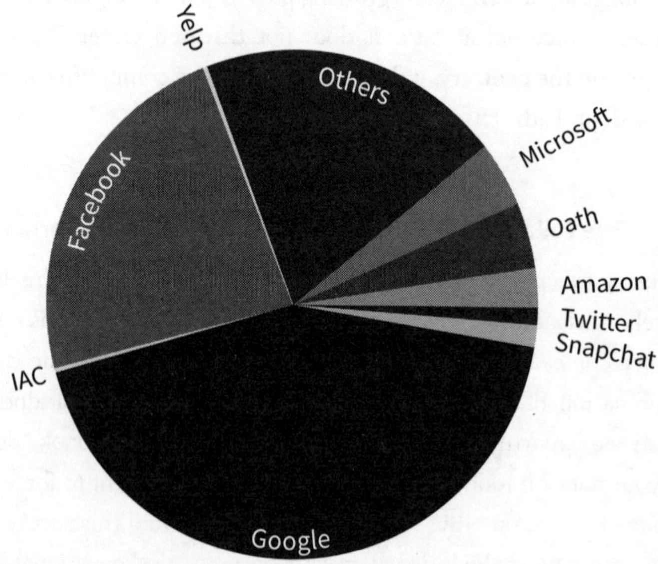

Source: Dipayan Ghosh and Ben Scott, *Digital Deceit II: A Policy Agenda to Fight Disinformation on the Internet*, September 2018 (www.newamerica.org/public-interest-technology/reports/digital-deceit-ii/competition/).

presence in the German social media market. The firm has 32 million users in Germany, representing 80 percent of the market share, just over double the threshold that by law invites scrutiny from European regulators. This presence practically forces the service on many German citizens, the Bundeskartellamt argues, and the firm's insistence that its users submit to the unchecked collection of personal information constitutes commercial abuse. The Bundeskartellamt's president, Andreas Mundt, concluded the following:

> Facebook will no longer be allowed to force its users to agree to the practically unrestricted collection and assigning of non-Facebook data to their Facebook user accounts. The combination of data sources substantially contributed to the fact that Facebook was able to build a unique database for each individual user and thus gain market power. . . . The only choice the user has is either to accept the comprehensive combination of data or to refrain from using the social network. In such a difficult situation the user's choice cannot be referred to as voluntary consent.[25]

While Mundt's arguments are intellectually honest with respect to the commercial situation at hand, this case represents a startlingly new direction for antitrust regulation: it is the first clear acknowledgment backed by the force of regulatory action that the systematic encroachment on the data privacy of the individual and the individual's relative paucity of alternatives for service access are closely intertwined and should be approached together—as a matter of competition policy—wherever the regulatory and political opportunities exist.

Until Mundt's charges, the world had not witnessed such a clear tying together of regulatory powers in the maintenance of market competition and individual privacy rights. And while Mundt's progress has been stunted after Facebook's successful appeal of the judgment,[26] the Bundeskartellamt's inquiry represents a serious step forward—and one that the U.S. regulatory community should give thought to as we consider an American approach to regulating internet competition.

Riding the Network Effect

When the internet was originally set on its path to commercialization more than three decades ago, it represented an entirely new communication medium and set of technologies that engendered an explosion of new industry and technological advancement.[27] But while attempts were made to apply a stringent regulatory regime to the development, installation, and operation of the physical network through the Telecommunications Act of 1996, no corresponding U.S. regulatory regime of any significance was developed to address the digital services that operated over the physical network.[28]

Some argue this is largely because common perceptions in the 1990s and 2000s held that competition for "over-the-top" (OTT) services offered via the internet was, if anything, fierce. The dot-com bust,[29] the variety of competing internet search services,[30] the persistent wane of various once-powerful internet brands,[31] and the dynamism and variety of new content available over the internet appeared to affirm these broad attitudes.[32] And while household names—Ask Jeeves, AltaVista, Yahoo, and GeoCities, to name a few—did exist in the mid-1990s, none of these were an apparent industry-consolidating machine, as Google, Facebook, and Amazon have become. None controlled an industry that amounted to a significant portion of the American economy overall, nor did any rival the economic capacity of major nations.[33] As such, none appeared unassailable in their commercial growth—and many were subsequently thrashed by the dot-com bust. In addition, the tools and technologies that have enabled the systematic economic exploitation of the individual consumer, including the capacity to store boundless quantities of information and analyze it using advanced learning techniques in real time, either did not exist or were cost prohibitive.

But the situation has degenerated further. As Amazon, Facebook, Google, and other giants of the American technology industry—including Apple and Microsoft—have beaten the competition to develop dominating products and services, each has grown to possess a stranglehold over its respective consumer market. Google controls the American markets for e-mail,[34] internet search,[35] and open mobile-operating systems,[36] Amazon controls e-commerce,[37] and Facebook controls social media[38] and internet-based text messaging.[39] Notably, each of these services originally benefited from the prevalence of the network effect, a basic feature of the consumer internet: the phenomenon that a service becomes increasingly valuable as more and more people use it.

WhatsApp, for instance, has grown to become one of the most popular consumer internet services in the world in part because it is the text-messaging service to which much of the world has gravitated. Amazon's e-commerce platform is so powerful as a shopping service for customers because sellers converge on Amazon—a dynamic that, in turn, brings more customers to the platform, thus inducing a cyclical dependence that would be hard for a competing shopping platform to break into. Google strikes strategic commercial relationships with producers of content, including music, television, movies, and influencer material, because by attracting them to YouTube over other streaming services, Google manages to maintain its historic levels of the online viewership market.

The phenomenon of the network effect is not new. Perhaps the simplest description of its impact is illustrated by Metcalfe's law, which depicts how the value of a network increases with a trajectory proportional to the square of the number of nodes in the network:[40]

$$Y \sim N(N - 1)/2$$

Meanwhile, the related Reed's law makes an even stronger assertion: the value of the network scales exponentially with the number of nodes:

$$2^N - N - 1$$

Some would suggest the Reed's law is more appropriate for the circumstance of the consumer internet, which can feature sub-groups.

As Dean Eckles, René Kizilcec, and Eytan Bakshy have shown, there are tremendously powerful network effects in digital media contexts, including social media platforms that may influence user behaviors depending on feedback they receive from their peers on the network.[41] So important is such achievement of the network effect—which can drive network-wide engagement over digital platforms like Facebook—that the venture-capital industry demands evidence of it before investing in internet-based start-ups. The corresponding investment culture preached in Silicon Valley has become clear to consumer internet entrepreneurs: chase users aggressively, using whatever means possible, before placing any constraints on your business model that have to do with protecting the public's interest. This stunning display of opportunism is typical of a consumer internet industry in which many seemingly established firms fail to make money until many years after they receive their first seed investments.[42]

Regrettably, the formation of such networks has also increased our risk of exposure to disinformation and induced the development of an insidious propaganda feedback loop in conservative media (an effect that has been analyzed at length in the excellent book *Network Propaganda*, by Yochai Benkler, Robert Faris, and Hal Roberts of Harvard University).[43] And what pushes many users of social media even further into their respective filter bubbles is the propensity for platforms to addict users. For many of us, the use of social media is like a game that has no end. In this context there is no such concept as economic satiation whereby the more you consume a typical good, such as espresso on a Sunday morning at a neighborhood coffeeshop, the less you will be willing to give up to get more of that espresso that morning. This is not coffee, or Belgian tripel, or Ray-Ban sunglasses. This is a novel currency. The platforms are dealing in our attention by feeding us material served up in a sequence engineered to engage our intellect; there is no diminishing marginal utility in the digital space so long as we are engaged by the content pathway we are subjected to by the platform. The opportunity cost—our time, which translates to hours we could have logged, or creative faculties we have expended, or the chance to spend time with friends and family—is vast. But we do not feel it.

Notably, at least one study, from 2019, suggests that ceasing use of Facebook, should one find a way to escape its addicting draw, can significantly improve a user's perceived well-being. Hunt Alcott and colleagues found that deactivation of Facebook led users to have substantially improved subjective well-being scores—upticks in well-being that on average were 25 percent to 40 percent as effective as a more traditional positive psychological intervention such as therapy or group training.[44] The deleterious mental effects of social media use appear then to be a reality and not an exaggeration. Nevertheless, as Facebook continues to expand into new projects that explore such new economic dimensions for the company as virtual reality with Oculus (yet another acquisition), cryptocurrencies with Libra, and others, it is actively positioning itself to secure its economic power even further.

Libra in particular should cause great concern for policymakers and the public. Here we have a private firm that has unilaterally decided, largely ignoring the concerns of the financial regulatory community, to create a new permissioned blockchain digital currency. Libra claims it will bring personal finance to the masses, including the poor around the world who stand to gain access to a digital wallet and the capacity to transfer currency without a bank account by creating "more access to better, cheaper, and open financial services—no matter who you are, where you live, what you do, or how much you have." We will see whether such charity comes to pass. There are many other cryptocurrencies—permissioned and permissionless—that already exist and that could also be used by people who are newly entering the fold of global commerce. But should the Libra project succeed, it will affirm Facebook's position in the space of personal finance and electronic transactions—and in the process tack a powerful new layer of economic value atop the company's universe of digital networks, solidifying its position in the technology industry ever further. And that is the company's true purpose.

The Systematic Concentration of Power in Digital Markets

We can conclude from the foregoing that the media ecosystem is simply undergoing a predictable evolution. Each year there are more and more eyeballs to be claimed: the global population is still expanding fast, especially in the developing world, where digital media dominates. Meanwhile, where there was more attention on television and traditional media in past decades, there is more attention on Amazon, Facebook, and Google today. The vast market for our collective attention and data is the target of these three firms, and they have successfully grasped their fingers around our eyeballs at the exclusion of the competition.

Where does this leave us? It would be misleading to conclude that competition is stiffening in digital advertising; that would suggest that Facebook, Google, and the potential Amazon five years from now will struggle to maintain their expected growth trajectories in digital advertising in terms of absolute revenues. This is unlikely to happen unless there is an overall stagnation in the market for consumer internet services and the ongoing harvest of customer attention and personal information. Instead, we might expect Okishio's theorem regarding the relationship between profits and competition to apply: as the firms continue to increment the productivity of their algorithmic decisionmaking, they will increase their profit shares in the absence of governmental regulations or public pressures that compel them to hire ever-more regulator-appeasing staffers, such as Facebook's legions of content moderators.[45] At their core, the platform firms' collective ad space is only as valuable as the size of the audiences they command in combination with their respective audiences' propensity to spend dollars; these three firms have each secured that valuable audience in their own way. In fact, we could see these three firms become more powerful over the next ten years and beyond, should the global regulatory community fail to take commensurate actions to curb their growing power. Perhaps this is indeed the most likely scenario.

This argument revisits our conceptualization of the business model that prevails across the consumer internet industry, whereby firms focus first on the creation and upkeep of compelling platforms like Facebook's news feed and Messenger, Instagram, Amazon, Google Search, or YouTube. These platforms, each of which has benefited massively from the establishment of the network effect, are so popular and pervasive that even for nonusers, it is hard to entirely avoid them, whether in terms of active use or simply in terms of surveillance. The second pillar of the business model is the unchecked collection of personal information on the individual through those services—collection that empowers these firms to develop and maintain robust and detailed behavioral profiles on individual users. These profiles are developed based on inferences made in real time about the individual using the amalgamation of data that firms such as Amazon, Google, and Facebook collect, including on-platform engagement data, browsing history, social graphs, location information, biometric information, and information purchased from data brokers, among other sources. And the third pillar is the development and ongoing refinement of tremendously opaque and increasingly sophisticated artificial-intelligence systems designed to do two things: curate social feeds and target advertisements at the individual.

All consumer internet firms follow this basic structure to monetize our attention and data. A key feature of this business model, consistent across the firms that implement it, is that the only element of it that directly contributes to corporate revenues—the targeting of digital ads on behalf of marketing clients—is a relatively small piece of the proposition that these firms offer to the market. They must practice the other elements with great effectiveness before they can engineer a proprietary platform for digital advertising. Correspondingly, this is precisely what we have seen in the case of Facebook, Google, and Amazon: each generally established indomitable leadership in its respective consumer internet market before seriously entering the digital advertising market.

With their success in dominating the market and pushing through

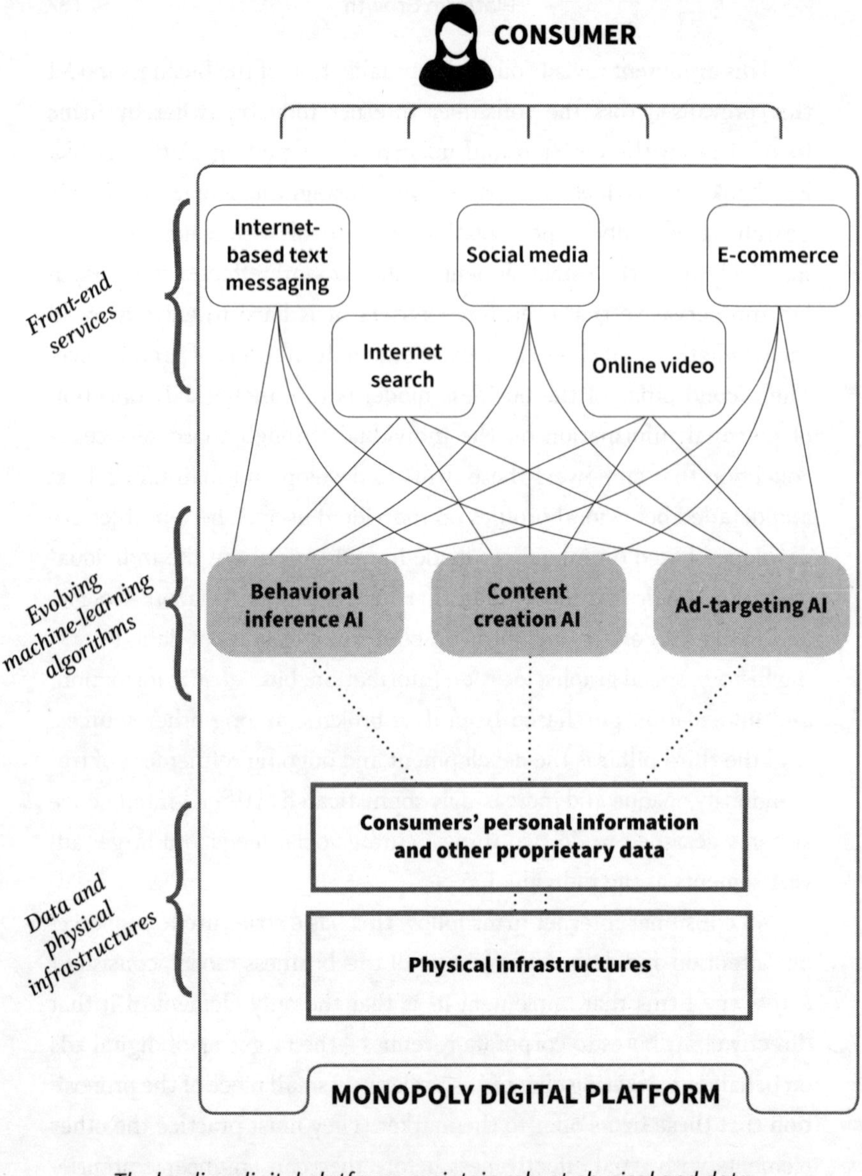

The leading digital platforms maintain robust physical and data infrastructures, implement them to develop sophisticated algorithms designed to engage consumers, and apply the resulting trained artificial intelligence systems to offer compelling services to end consumers, whose data and attention are raked at monopolistic rates and monetized in the face of digital marketers, whose objective is to influence consumer actions in the marketplace.

the new technological paradigm concerning the media and information landscape has come the added benefit of securing the future of these firms. As suggested by the Lindy effect, the technologies underlying the monopoly digital platforms—and the companies that operate them—are becoming less and less likely to die off anytime in the foreseeable future as we move forward in time.[46] That is to say, the life expectancy of the companies and their marketed technologies increases day by day, and as they become ever more entrenched in the media space, they become more known to marketers and consumers as the forum to which they should bring their business and attention. And there is no law of the handicap of a head start that applies here; these firms moved early in the commercialization of the internet and have uniformly secured vertically integrated market power throughout the internet-media technology stack that without governmental intervention will perpetually serve as the ultimate natural barrier to competitive market entry.

Smaller Rivals in the Consumer Internet Industry

The leading internet platform firms have had an extended period of low public perception, which may have helped smaller firms that possess some market share. This takes nothing away from the enterprising nature of firms such as Snap, whose principal social media service, Snapchat, brings in the vast majority of its advertising revenues. But we must see its recent near doubling of market share in digital advertising not as Facebook's competitive bugbear but as no more than the result of a vigorous strategy of growth and entrepreneurship with data and novel practices and new offerings for advertisers, an *annus mirabilis* that has most likely occurred because the larger internet firms have had a couple of bad years.

For instance, Facebook, in the wake of the Cambridge Analytica revelations, was practically compelled by public sentiment to shutter Partner Categories, its service that enabled data brokers to inject their data into the Facebook advertising platform at the behest of advertisers.[47] This was the tool that allowed, for instance, Audi to target an ad campaign for a new sports car at persons who had walked into a BMW or Mercedes dealership any time in the past few months. Of the four major American data brokers, it is Oracle that collects this information on potential sports car customers. Audi could target walk-ins to the other manufacturers' dealers by selecting that audience through Facebook's Partner Categories interface, and Oracle would take a cut from each dollar spent in disseminating marketing messages to that segment over Facebook's platforms.

On the heels of the Cambridge Analytica disclosures, Facebook quietly announced in a three-line press release that it would shut the service down—most likely to avoid further public scrutiny over the questionable and opaque transfer of millions of people's personal data between various companies at the click of a button, in combination with the fact that Facebook itself could recoup much of the lost business by investing more in its own analytical technologies designed to infer consumer behaviors and preferences. Within days, Snap announced that it would commence a similar targeting capacity with brokered third-party data, which it was able to do only because scrutiny over the firm had all but ceased while the public was attuned to the harms perpetrated by others in the industry.[48]

The Coming of Age of the Modern Natural Monopolies

There are two kinds of monopoly: those that are natural and those that are not. The latter is characterized by the practice of artificially sealing off access to markets by raising barriers to entry, among other anticompetitive practices. But one can imagine a situation where the firm in question could survive even if it had a rival competing in the same market. The home appliance market presents a good example. As the Open Markets Institute notes, "Whirlpool's takeover of Maytag in 2006 gave it control of 50 to 80 percent of U.S. sales of washing machines, dryers, and dishwashers and a very strong position in refrigerators."[49]

We could easily imagine Whirlpool having a competitor in this market; namely, another appliance wholesaler. Its market monopolization is characterized by a strategy of aggressive acquisition.

In a natural monopoly, the monopolist could not have a viable competitor because it will never make economic sense for a competitor to enter the market. Three leading examples of natural monopoly industries come to mind from U.S. history. The first is the railroad industry. Typically, it only ever makes sense to have one set of tracks in a rail network. Though installing a second set could induce competition on the incumbent, the capital expenditure involved in developing the second rail network would be too great to bear. The second example is the electricity grid. Here, again, it makes economic sense for society to invest in just one network. Why should citizens and consumers foot the bill for the construction of two sets of substations, two transmission towers, and two distribution networks all standing next to each other? And the third is telecommunications, an industry that, like electricity and rail, features a physical network that necessitates high front-end capital expenditure for build-out with high subsequent operating costs. Once we have one, it makes little economic sense, even in a capitalistic economy, for anyone to invest in a second.

The leading consumer internet firms—Facebook, Google, and Amazon—are a new kind of natural monopoly. To be sure, this is a controversial assertion; whether the leading consumer internet firms are monopolies at all, let alone natural monopolies, remains very much an open question. But I believe it to be the case: there can be no second search company, e-commerce firm, or social media conglomerate that will be able to compete with these three firms. There might be smaller competitors, such as Snap, Bing, and Walmart's e-commerce platform, but they will remain far behind the leaders in perpetuity. The lion's share of the consumer market will go to Facebook, Google, and Amazon.

There are a few hallmarks of the classic natural monopolies of the American industry, and all are true of today's consumer internet industry. The first is the networked nature of the firms in question. The telecommunications, railroad, and electricity industries are all signified by their physical networked nature; they require that holes be dug and ditches and trees be cleared to run the relevant lines through the ground. Each has nodes of activity, whether they are substations, railway stations, or terminal hubs. The consumer internet industry is also highly networked, albeit in much more fluid and transient form, as users come and go. The consumer internet thereby benefits from the network effect, the feature by which the value of the firm increases as more and more people choose to join the network. We will only ever wish to expend our time and energy on one primary messaging service, one social media service, one search platform, one e-commerce platform. Why have two? The economic cost of instituting the second—of having all your friends leave Messenger and join a new texting service, or of having all our phones transition to a new search service—is far too great to allow for commercial success. We will always gravitate toward the one service, given how utility-like the provision of social media service (or e-mail, or e-commerce) has become. It is technological somnambulism at an industrial scale.[50]

The second natural-monopoly trait is the high capital expenditure for initial build-out of the network in question. The costs of building out telecommunications, rail, and electric networks are clear. There are similar costs in building out social media networks, search networks, and e-commerce platforms. These outlays range from the physical to the digital: server farms, connectivity to the network, commercial arrangements with telecommunications firms and content delivery networks, build-out of the end-consumer platform, development of the algorithms to support content curation and ad targeting, and establishment of relationships with marketers. While all of the consumer internet platforms started out small, over the next several years each expended billions on building out the infrastructure needed to support their services.

The third hallmark of a natural monopoly is the vertical integration of industries. This is exemplified by the traditional physical networks. Before its deregulation in the 1990s, the electricity industry, for instance, was integrated at all of its levels from top to bottom: power generation, transmission, and distribution. We can observe a similar theme in the case of firms such as Facebook, which have vertically integrated the entire consumer pathway, from providing the core social media service, to tracking customers along their pathways over the service, and finally to targeting customers over the service using their personal data. This vertical integration has become so intensified in the case of Facebook, Google, and Amazon that each represents a walled garden; the judgments they make about our character and personality, and the ways they use that knowledge, are known only to them. In this world there is no room for investment-specific technological progress for the upstart that might wish to compete with the likes of these monopolies.[51]

Regrettably, the leading consumer internet firms' market power as established through their natural-monopoly status has furthermore invited each to systematically engage in a seemingly endless list of economic injustices against the American consumer. Exclusive dealing, whereby the consumer internet firm forces other firms to use its own services or buy its own products; refusal to deal, whereby the consumer internet firms pick and choose their business partners on an unfair basis; and tying, whereby firms bundle their offerings together even when the consumer market does not demand it—these are all examples of incidents in which consumer internet firms have perpetrated harms to vertical supplier-distributor relationships.[52]

And the list goes on, including the predatory poaching of countless firms that could have otherwise posed a competitive threat if their acquisitions had been blocked over competition concerns; predatory pricing, whereby the established consumer internet firms have undercut their rivals in attempts to throw them out of the consumer market early; the monopoly rates at which the combination of our data and attention are gathered and monetized in the digital advertising market; and the terrible quality of service some consumers have experienced, particularly as novel negative externalities have emerged out of the very makeup of consumer internet firms in recent years. These harms pertain to artificial restrictions promoted by the consumer internet firms to the end of closing out horizontal competition. The Organisation for Economic Co-operation and Development (OECD) offers a list of potential consumer harms that can result from a lack of competition in markets, each of which feels like a palpable reality with respect to the consumer internet: higher prices, reduced output, less consumer choice, loss of economic efficiency, misallocation of resources, and reduced quality of output.[53] We have not seen the sailing ship effect in the consumer internet sector even though new technologies have been introduced to the market in recent years; there has been no new firm that has forced new innovation out of the incumbent monopolies. The need to pursue real innovation falls away when new players can simply be acquired or copied at limited cost and without scrutiny.[54]

Meanwhile, the collection of data pertaining to the individual and the injection, filtering, and ordering of content for the individual through the platform service and beyond—including such external third-party contexts as those promoted by Facebook's Audience Network, Google's Display Network, or Twitter's Audience Platform—has uprooted the notion of individual privacy. The commercial regime that underpins the consumer internet engenders an economic logic that encourages the relegation of consumer privacy as a mere afterthought for the purposes of the firm hosting the advertising platform. As many have commented, this market activity encourages a fundamental erosion of the individual's privacy and autonomy—elements over which rival firms could compete should they be able to overcome the barriers to entry erected by the sector's current dominators.[55]

Furthermore, the platform firms' dependence on data for content curation and ad targeting has directly encouraged diverse negative externalities that have shaken our democracy. At issue is the fact that the business models prevalent in the sector encourage ongoing use of the relevant platforms by predicting the content that algorithms have inferred are most likely to engage the consumer for the longest period of time, no matter what the nature of that engagement might be. Mean-

Facebook Default Visibility Settings Over Time

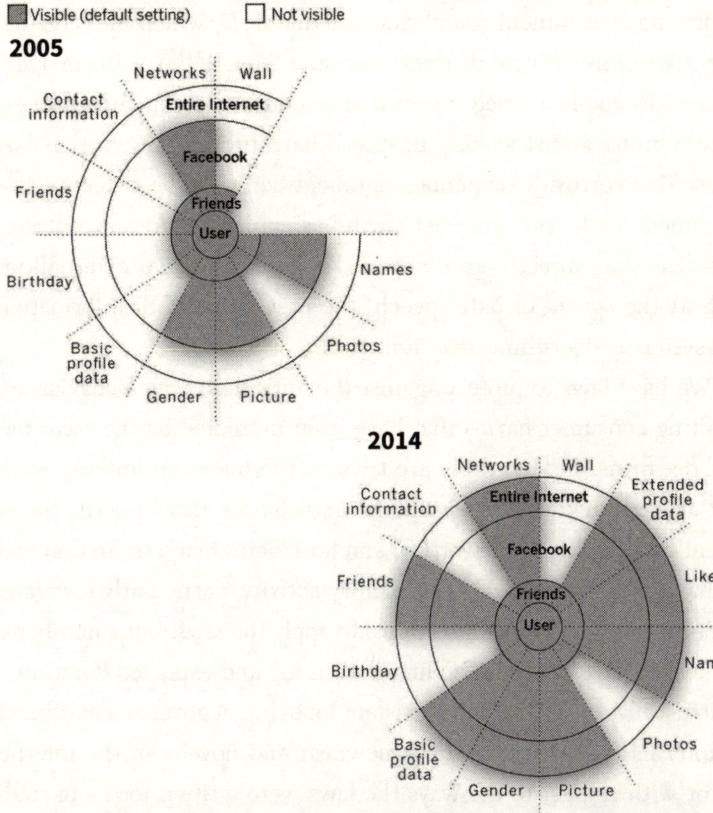

Facebook has an economic interest in exposing many aspects of user privacy, including visibility of the user's content, such as who the user's friends are and what pages the user has liked. By maximizing such sharing and visibility of content to the greatest extent possible, firms can in turn maximize users' engagement among each other, thus attracting the user population to stay on Facebook more, contribute more data to the network, generate more ad space, and increase well-informed ad-targeting opportunities.

Source: Alessandro Acquisti et al. (2018), "Privacy and Human Behavior in the Age of Information," *Science*, vol. 347, no. 6221.

while, the platforms have traditionally allowed any third-party advertiser to initiate ad campaigns, contingent only on its ability to pay, thus instituting an unflinching and vicious dynamic by which the economic incentives of the content disseminators and those of the platform operator are absolutely aligned—even if the content disseminator happens to be a propagandist looking to spread harmful content, such as fake news. This corrosive economic alignment between the two entities—combined with the internet firms' singular profit-maximization objective—has directly encouraged damaging negative externalities, such as the spread of hate speech, the prevalence of disinformation, and systemic algorithmic discrimination.

We have laws to protect against the anticompetitive behavior and resulting consumer harms that have been promoted by the consumer internet firms. In fact, there are laws on the books around the world that are designed to address any firm's behavior that imperils the efficient functioning of the vertical and horizontal markets. To that end, we have seen a great deal of regulatory activity, particularly instigated by European regulators, in attempts to apply the laws. But a number of challenges have arisen, including the timing and expected duration of antitrust investigations, the persistent lobbying of government officials by internet companies, and the newness and novelty of the internet sector with respect to the ways the laws were written for more traditional industries.

We have to invite a whole new form of regulatory scrutiny over the sector: we need to start treating the consumer internet giants like natural monopolies until and unless they stop exhibiting the features we typically associate with natural monopolies. Until we take this critical step, we will continue to see decidedly lame judgments, such as the Federal Trade Commission's $5 billion fine of Facebook in 2019—an intervention that may appear minatory but that will ultimately do little to halt the company's violation of the public interest.[56]

The Desired Market Outcomes of Reviving the Dynamism and Vibrancy of the Internet

American competition policy is chiefly concerned with two principal objectives: to benefit consumer welfare and to protect the dynamism of markets. Developing jurisprudence in the United States over several decades has defined promoting consumer welfare primarily as ensuring that the prices consumers pay in exchange for goods and services are fair.[57] Dynamic markets prevent stagnation and ensure that firms that rise to the top of American industry do so owing to the merits of their competitive strategy. As an Obama-era Council of Economic Advisers report notes, "When firms attempt to increase their profits through anticompetitive means—colluding with rivals, purchasing competitors, erecting barriers to entry to insulate their incumbency from competition, or other actions—society suffers."[58] We all want to see a regular cadence of new innovations hit the consumer marketplace. And we want to persist in driving the national and global economy forward.

Despite the capitalistic design of our national economy, the United States has always placed the integrity of democracy over the commercial interests of the markets. The break-ups and pro-competition policy resolutions to address J. P. Morgan, Standard Oil, AT&T, Microsoft, and many others serve as critical examples. In each case, the inordinate commercial presence developed by a single firm was deemed to constitute monopoly leveraging, and efforts were undertaken to encourage greater competition. Indeed, the passage of the three main laws that define the U.S. governmental powers in protecting competition—the Sherman Antitrust Act, the Federal Trade Commission Act, and the Clayton Antitrust Act—were put forward to protect consumers from anticompetitive harms.

These arguments can now be made of the American technology industry; federal action must be taken to protect the American consumer. But over the past forty years, the courts have defined monopoly more narrowly. The prevailing view is that a firm should not be subject to corrective action unless it has been established that its operation leads directly to consumer harm (that is, in the practice of modern antitrust enforcement, price hikes).[59] Many hold that this point of view has diminished market competition—in particular, and with respect to the consumer internet, that it has raised barriers to entry, worked against the cultivation of a level playing field, reduced overall market dynamism in a once-vibrant industry, and diminished quality of service, as evidenced by the negative externalities present on these platforms. Furthermore, scholars have argued that Facebook, Google, and Amazon charge monopoly rates to consumers.[60] A federal policy on this matter should work against these harmful dynamics.

Google, Facebook, and Amazon have collectively exploited the end consumer and impeded the vibrancy and dynamism of the internet. And although competition law in the American tradition "does not regard as illegal the mere possession of monopoly power where it is the product of superior skill, foresight, or industry," since such monopoly status need not necessarily imply consumer harm,[61] these three behemoths have each ridden the network effect and undertaken tactics to secure, wield, and expand their power over their respective market silos.[62]

The consumer internet's modus operandi must prompt regulatory scrutiny. Each of these firms has now taken its respective market and benefited from established network effects while simultaneously making strategic acquisitions to block off the capacity for competition. American consumers now face a harsh reality: the consumer internet market is controlled by a few inordinately large firms that, if left untouched by the government, will continue to perpetrate harms against the American public given that they are allowed to do what they will with our data and the competitive circumstances in the markets at large. This is wholly unfair to consumers. We deserve competitive markets that elevate the companies that innovate the best—and in doing so serve the customer first.

FIVE

A New Social Contract

The tide of public sentiment is closing in on Silicon Valley. We are witnessing market failure at a grand scale. As the leading internet firms have monopolized their subsectors, individual citizen and consumer interests have fallen by the wayside. The Cambridge Analytica revelations, frequent disclosures about privacy and security breaches, and shocking policy decisions implicating democratic processes have become second nature to the internet sector—and have rightly forced our attention fully on the inner workings of Facebook, Google, and Amazon. The *vox populi* stands squarely against the industry's interests.

As with any market failure, government must eventually intervene. These firms have inordinate market power in the end-consumer market. Facebook, for example, has monopolies in traditional social media and internet-based text messaging, Amazon has a monopoly in e-commerce, and Google has monopolies in online video, e-mail, and search. These firms are able to rake currency in the form of attention and personal data on one side of the market and charge monopoly rents for it on the other side. They have breached privacy, tampered with markets, lowered our expectations of the open internet, and shaken the very foundations of our democracy.

This hegemony over the market has been shown to trample the public interest. The industry's disincentive in protecting the public from such negative externalities as disinformation is a mere symptom of its unwillingness to compromise its highly profitable business model along with its use of social power to protect the business model from regulation through influence over policymakers. Installing sleeping policemen to place soft checks on the industry's practices will no longer resolve society's problems at this present juncture. What the public now needs is a novel, comprehensive regulatory regime that can effectively rebalance the distribution of power among the industry, government, and citizen—a digital social contract.

Noting the viciousness of unregulated capitalism through affordances of radical property rights, the philosopher Jean-Jacques Rousseau suggested that "if we have a few rich and powerful men on the pinnacle of fortune and grandeur, while the crowd grovels in want and obscurity, it is because the former prize what they enjoy only in so far as others are destitute of it; and because, without changing their condition, they would cease to be happy the moment the people ceased to be wretched."[1] We are at this stage of political evolution in the context of the internet sector.

Never can one pick a rose without feeling the prick. I, however, do not contend that we should abolish the industry's targeted advertising business model entirely. Rather, I suggest the development of a novel regulatory regime that effectively responds to the overextensions of the economic logic residing behind the consumer internet. This new regulatory regime must include measures that can effectively hold the industry accountable—new mechanisms ranging from radical economic redistribution policies to automatic stabilizers of economic power. If the industry has converged on the practice of uninhibited data collection with the end goal of refining behavioral profiling, we need consumer privacy. If the industry has largely obscured its core content practices from the public and the regulatory community at the expense of marginalized classes, we need transparency into its corrosive platforms. And if the industry has systemically abused the open markets through engagement in anticompetitive practices, we need robust reinstatement of market competition.

Free market capitalism is the principal hallmark of the American approach to economic design. Open capitalism defines the operation of not only our national economy but indeed our political and social systems. But the American government has never hesitated to strike the industry down when its practices have imperiled the nation's commitment to democracy.

This is the very situation we now find ourselves in with respect to the internet. To curb Silicon Valley's capitalistic overreach, the federal government must formulate a cohesive policy response aimed at protecting our privacy from the internet sector, promoting market competition in the industry, and providing us with transparency into its practices. And this regulatory regime can only achieve maximal impact if all of these reforms are passed together, thus treating the current makeup of the industry's economic logic with a comprehensive regulatory framework to meet the industry head-on.

Privacy

The political and practical creation and passage of a rights-based digital agenda will be a major challenge for policymakers in the years ahead, and the core of that agenda must be focused on consumer privacy, perhaps the most critical of reforms.

Appropriate privacy protections can do much to blunt the cunning of precision propaganda as disseminated by the Russian disinformation operators, the spread of hateful conduct online, and the drive toward online radicalization. In each case, it is the consumer internet platform's predilection to algorithmically analyze the behavioral profile of users—including their likes, dislikes, preferences, beliefs, interests, and routines—that leads them down the dark rabbit hole of these negative externalities, sometimes even at the direction of an advertising client, such as a disinformation operator working through a shell organization. Without such a detailed understanding of their behavioral profiles, platforms would be unable to direct customers down a path of hatred, lies, and racism, whereby content recommendations at the determination of the platform lead the user to radical content—as we have repeatedly seen in the case of YouTube.

The reforms necessary to promote strong privacy protection are quite clear. There is a well-established tradition of privacy protection in certain jurisdictions around the world. Indeed, a strong approach is already on the books—a version of the European General Data Protection Regulation (GDPR) packaged with strong regulatory enforcement capacities—one that we could adopt or take as a model.

Privacy protection is the one reform that the industry's henchmen will fight tooth and nail. Only in special circumstances has it passed. In formulating the new regulations, the European Union sought to bring its fundamental right to privacy, affirmed in the 1996 Data Protection Directive, into alignment with the requirements of the contemporary digital economy. The political impetus was there. Europeans took to the streets in protest following whistle-blower Edward Snowden's 2013 disclosure of American coordination with Silicon Valley firms and their engagement in mass global surveillance operations. In California, reformist Alastair Mactaggart smartly threatened a stringent privacy bill while coordinating a reasoned negotiation with the industry and the legislature; only the threat of the former, fueled by a quirk of the legislative function in California, could have activated the latter.

We have seen great courage from some of our politicians on the specific matter of technology regulation from both the right and the left, including Republican senators Josh Hawley and Lindsey Graham and Democratic senators Elizabeth Warren and Mark Warner. But this will need to spread across the board in both the House and the Senate, as well as across the aisle. Economic regulation is typically a partisan concern, and politicians will need to make the case in clear terms that reforms are needed. Their constituents, in turn, should have an open mind to understand the issues at the heart of the industry so that they too can formulate their own opinions.

That some jurisdictions have passed and already enforced strong privacy regimes means that much of the intellectual work has already been done. We require a rights-based regime—a privacy regulatory standard that places power in the hands of the individual and affords consumers the power to control their data. This will naturally be challenged by the industry. Their margins are historically high despite low innovation across the sector today because of the systemic exploitation of the consumer. We must break that hegemony. Privacy is the place to start.

Consent, Choice, and Control

Modern privacy protection today in practice cannot focus on dictating whether firms can collect various kinds of personal information. The collection and processing of personal data are so ubiquitous and continual that forbidding the practices would be far too complicated an exercise to legislate or to enforce.

Privacy laws must start with giving individual consumers—the users of such platforms as Facebook and Twitter—complete control over the access, use, and retention of their personal data. This means that, as a default, firms must be compelled to acquire users' explicit consent before collecting or processing their data. Consent should be offered on an opt-in basis, whereby consumers explicitly permit the firm to collect and process their data. Nonexistent or cursory requests for our permission will no longer suffice; we have to begin forcing rigor on an industry that has subjected the public to one harm after another on the basis of unchecked collection of our data. The GDPR contends that opt-in consent must be "freely given, specific, informed and unambiguous" by means of a "statement or a clear affirmative action"—in essence, an engagement with the firm that is clearly about offering consent, with users having complete access to any information pertaining to the firm's eventual activities with their data. Hitting the bright green "Accept" button next to a pre-checked box will not suffice. This should be a fair economic exchange, not a mass hoodwinking of an uninformed consumer market.

Should consumers wish to withdraw consent, they should be able to do so with immediate effect. Our data are an extension of our individuality; should we no longer wish to lease our data out to a profit-seeking firm, we should have the right to do so on a moment's notice. As the GDPR notes, a consumer's consent "shall be as easy to withdraw . . . as to give." The consent framework should apply to both the collection and processing of our data. Courtesy of the GDPR, European citizens have the right to forbid consumer internet firms from processing their data. The right to object to commercial use of our data—commercial use focused on, say, developing inferences about us without our full understanding and then letting any advertiser willing and able to pay for access to those inferences target us with ads—is an option all internet consumers around the world should have.

There must be specific and direct consideration in regard to cookies—those invisible trackers used across the web—particularly when they are used for commercial purposes that the user does not sign up for or expect. This restriction would apply to the vast majority of cookies that are implemented on the internet, including those used principally by the consumer internet firms to maintain their hegemonic control of the media ecosystem. Firms should be forced to collect additional specific consent from users to implement cookies in the user's browsing experience.

Technical Functionality Requirements

Consumer internet firms offer the user a core service, whether the service is centered on provision of social media services, e-mail, or text messaging. Presumably, firms have to collect some data to enable the full functioning of these services. A private e-mail service, for instance, may need to know your name and may additionally wish to collect your phone number so that it can apply two-factor authentication to your account. It might also wish to have access to your e-mail contact book to enable you to easily look up your friends, family, and colleagues as you compose a message. It could be said that these are categories of data that the service requires to maintain technical functionality of its service. What it does not need to do, however, is read your messages and attempt to infer your profession, education level, income level, ethnicity, psychological tendencies, and political leanings. Nonetheless, consumer internet firms do this across most of their services, most likely with little knowledge of the user throughout that data collection. This has to end. To return power to the consumer's hands, governments should stipulate that data collection and maintenance beyond what are minimally necessary for technical functionality of service cannot be undertaken without a consumer's explicit opt-in consent.

The Rights to Access and to Be Informed

Transparency is key. For too long, internet firms have gotten away with hiding from users how much data they collect, what tools and technologies they use to collect data, and the reasons behind their collection of the data. We need governments to mandate that internet firms—and any others that intentionally obscure the real reasons for their data collection from the typically unassuming and uninformed consumer—must detail their data-collection and data-processing practices in clear terms.

Public frustration over obscurantism on the part of the consumer internet sector prompted the California Online Privacy Protection Act of 2003. And because any firm that has any California residents as users must abide by that law or be thrown out of the California consumer market, the law has become a global standard. However, the law has some gaps that need updating for the modern context. In particular, while requiring a detailed version of consent is pertinent, firms should also be required to offer a version that is clearly understandable to the layperson.

In addition, consumers must be afforded a clear right of access to the data maintained on them by the firm—and it must be easily accessible. We need to know what the firms have on us so that we can make reasonable determinations about what we are getting for the data and time that we are giving up to the firm. The rights to access offered to European citizens through the GDPR are clearly not enough; they will need to be enforced far more effectively. As Carl Miller, a researcher at Demos, the London-based public policy think tank, illustrated one year after the GDPR came into force, companies are not well positioned or incentivized to provide access to a user's personal data in the absence of enforcement of access rules.[2]

The Right to Correct Inaccurate Records

Consumer internet firms are constantly collecting data on us. Some of that information is relatively clear to us. For instance, it may be associated with data that we had to enter to gain access to the service (such as our name and e-mail address), or perhaps with our direct engagements on Google Search. But some of the data that firms have on us are simply inaccurate. Google may have obtained data from some third party that made an arrangement to sell the data, and perhaps that information goes into a behavioral profile designed to model our personality.

As you may have noticed, much of the information that data brokers possess on you might be inaccurate. They might have the wrong Jane Doe, or they may have had ineffective data-collection practices or low-quality data partners, but at the end of the day all you care about is that some of your data are wrong—your age, or profession, or level of education, or car. Owing to long-standing transparency measures, some data brokers have started to allow consumers to see some of the information they possess on us. Acxiom's tools are perhaps the most visible example. But such tools need to be made much more readily available to consumers—by data brokers, internet firms, internet service providers, and beyond.

The GDPR affords a right to rectification, whereby the company has to correct any discrepancies within thirty days of notification by a user that his or her information is incorrect.

The Right to Be Forgotten

The central notion around the right to be forgotten is clear: consumers have a right to disappear from commercial data banks. The right to be forgotten dictates that consumers can request the deletion of data that are no longer needed for technical functionality of the service rendered by withdrawing consent or rejecting further processing of their data. In any such case, the consumer internet company would have to comply with the consumer's request. These are the standards stipulated in the GDPR, and a right to erasure has also been enshrined in the California online privacy protection law through a 2013 amendment. In a new regulatory regime, such stipulations should be applied to new laws around the world.

Notably, some, particularly in the industry, have raised concerns that affording such a right would impede free speech. I consider this a minor concern that has been blown to extravagant proportions by the industry in efforts to uproot the EU's policy. It is true that we must respect freedom of speech and freedom of information. There are some cases where this is critical: when the information is a matter of public record, or when maintaining its public availability reasonably advances the public interest, the data should not be deleted.[3]

While the line set for this standard will vary by jurisdiction, which might introduce some technical complications for the particular internet firms, we should also remember that the leading internet companies can easily apply those constraints should it be deemed necessary to their proper functioning.

Particularly Sensitive Data

Some forms of data are more sensitive than others. Indeed, this is precisely why the United States has privacy laws that apply only to certain sectors that the government has determined require a baseline of privacy protection for citizens. These sectoral laws include the Electronic Communications Privacy Act, which covers electronic data transmissions; the Gramm-Leach-Bliley Act, which applies to financial information; the Health Insurance Portability and Accountability Act, which protects healthcare data; the Children's Online Privacy Protection Act, which covers the data of minors age thirteen and younger; and the Family Educational Rights and Privacy Act, which protects educational data.

As jurisdictions around the world formulate local privacy laws, extra consideration should be given to any data generated in such sensitive contexts. Additional consents from the individual and more rigorous security measures should be considered for data pertaining to such contexts.

Furthermore, consumers should be afforded extra rights pertaining to certain categories of demographic data that can otherwise be used to discriminate to potentially harmful ends. The GDPR offers such protections for "special" data, which require "explicit" consent for data processing by the firm. The processing comprises "particularly sensitive [data] in relation to fundamental rights and freedoms," including data "revealing racial or ethnic origin, political opinions, religious or philosophical beliefs, or trade-union membership, and the processing of genetic data, biometric data for the purpose of uniquely identifying a natural person, [and] data concerning health or data concerning a natural person's sex life or sexual orientation." For such forms of highly sensitive data, consumers should have to explicitly consent before firms engage in processing data that are not necessary for technical functionality of service provision.

Coverage of Profiling Inferences

Your raw personal information, such as data associated with your on-platform engagement with Facebook's news feed or YouTube's search bar or data acquired through cookies and from data brokers, is tremendously valuable to the consumer internet giants. But it is valuable only because it is the information these firms use to develop inferences about who you are. Those inferences can come in many forms. The "Interests" section of your Facebook home page is a good example. These are the interests Facebook has inferred you have, from the Cleveland Browns to Bulleit Rye whiskey. Inferences might also be retained in more abstract form, perhaps as a set of conclusions an internet firm has drawn about you that aggregates all data that the firm has obtained and analyzes the data on an ongoing basis to develop a digitized approximation of the "real you."

It is important that privacy laws target these data—the inferences that the firms have drawn about us—in addition to the raw data that firms obtain. This is a somewhat controversial contention. Consumer internet firms have variously attempted to suggest that such stipulations would impede democratic freedoms and diminish their incentive to innovate in their use of data and to provide optimally effective products to consumers.

Despite the industry's weighty but flawed arguments, governments must realize that the raw data firms possess are simply less critical to the leading internet firms' commercial success than the inferences the firms develop about us. It is largely the inferences, not our raw data, that the firms use to target ads and curate our social content. If it is these inferences about us that firms are using to determine what we should or should not see in an era of rhizomatic data discrimination, consumers should have the right to see those inferences, correct them if desired, or delete them from company servers.

The GDPR covers a version of such inference development in what it calls "profiling" of the individual, which might entail analysis of "aspects concerning that natural person's performance at work, economic situation, health, personal preferences, interests, reliability, behaviour, location or movements." The GDPR notes that individuals should be made aware of the technical logic used in any profiling practices, as well as the underlying purpose of the profiling. The firm must stop profiling users who indicate that they want the firm to cease the profiling. These constitute strong and important protections for European citizens that should be extended to all internet consumers.

Data Anonymization

Firms typically maintain their data on us in unregulated formats. For example, most jurisdictions impose limited, if any, requirements that firms use particular types of encryption to ensure reasonable levels of data security. It is furthermore quite difficult for regulators to create highly detailed security standards that can be applied across the industry, since firms' nature and maturity are so variable across many firm-specific factors.

The GDPR applies an innovative approach to resolving the tension around user anonymity by providing incentives for firms to use something the regulation describes as pseudonymization, or "the processing of personal data in such a way that the data can no longer be attributed to a specific data subject without the use of additional information." That additional information should be stored somewhere separately and is subject to certain security measures. The GDPR notes that firms might be allowed to process pseudonymized data for purposes outside the original scope of its collection and to satisfy some of the GDPR's data-security requirements—important considerations for technology firms large and small.

I believe this remains the most promising governmental approach to fostering industry innovation in the data-security space we have so far seen. Legislatures and regulatory bodies should consider mechanisms to encourage use of such pseudonymization and anonymization techniques to protect the privacy and security of the consumer. In particular, firms should be encouraged to implement incrementally novel encryption techniques through provision of safe-harbor protections, as the Europeans have done.

Administrative and Operational Stipulations

There are a number of other detailed requirements that governments should consider establishing to protect consumer data. These include requiring that firms conduct periodic impact assessments pertaining to their data-protection practices; engage in robust data-breach notification campaigns whereby authorities, regulators, and consumers are promptly advised of a breach once firms become aware of it; impose and enforce strict and accountable agreements with any third-party technology vendors; and record any data-processing activities to facilitate potential future forensic review. The GDPR includes provisions pertaining to each of these.

We need a privacy regulation in the United States that rebalances the distribution of economic power between the digital platforms and their users. The situation today favors the industry; firms have complete unilateral authority to collect whatever data they wish and take any action with it that they desire so long as they do not breach previously stated agreements with users. That needs to change. The GDPR takes the most significant steps in arresting this subversive circumstance, steps that now legislators in the United States and around the world should too consider. And while the GDPR and like regulatory regimes might diminish many aspects of the programmatic advertising systems that the firms have developed, we must also recognize that there have been times throughout history when powerful industries have had to be curbed to protect the democratic interests of the nation. Our current circumstance is no different in the face of a Silicon Valley that looms over our political system and media regime.

Competition

There is a wide range of political perspectives on how competition policy should be applied to internet firms. Many have argued that, in response to the growing policy concerns associated with the competition among leading internet firms, the global regulatory community—particularly in the United States, where the developing jurisprudence has adopted a staunchly conservative approach—brings novel forms of antitrust regulation and enforcement to bear against the industry. Tim Wu and Barry Lynn advocate a shift from the relaxed and consumer harm–oriented mode of antitrust enforcement that has been taken up by the courts in recent years toward a return to the more progressive trust-busting approach associated with Justice Louis Brandeis, President Theodore Roosevelt, and President John F. Kennedy—calls that have been taken up by political leaders on both sides of the aisle, including Senators Elizabeth Warren and Lindsey Graham.[4]

Others have argued that the consumer-harm standard prevalent in the United States has the important virtue of giving antitrust regulators clear guidelines on judicial interpretations, which empowers other parts of the law and other enforcement regimes to act when necessary, obviates the critical problem of regulatory capture, and offers the business community credible certainty over what business practices are allowed and disallowed. Various scholars have advocated a considered approach. Carl Shapiro, for instance, has noted that there is limited evidence indicating the diminishment of social welfare caused by the leading internet firms, in efforts that perhaps echo those of Frank Easterbrook's earlier scholarship.[5] Jason Furman notes that regulators should not risk sacrificing the efficiencies of the leading internet firms by taking too aggressive an approach, as those efficiencies can benefit consumers.[6]

Still other scholars recommend a different kind of caution in reform movements, noting that novel enforcement cases that might seek to take advantage of a progressive interpretation of the still ambiguous Section 5 of the Federal Trade Commission Act should carefully consider the typically long timeline of the prosecution of monopoly cases, the tremendous resources required to pursue such prosecution, and the potential deleterious effects of antitrust enforcement in the long run.[7]

Considered according to the actual constituent markets they serve, I believe that Facebook, Google, and Amazon should be seen as out-and-out monopolists that have harmed the American economy in various ways—and have the potential to do much greater harm should their implicit power go uncurbed. Without some form of intervention, the situation will not change because presently these firms have protected their status and appear poised to continue that trajectory into the future absent the appropriate governmental interventions. The only future these firms see is one in which their lead in their respective markets is untouchable, which, as we have discussed elsewhere, begets a natural monopoly. They should therefore be subject to the regulations typically imposed on existing natural monopolies—the telecommunications, railroad, and electricity industries.

A progressive approach to competition policy is needed to break the market control exhibited by these three firms—the theory of the case being that it is the dominance of these firms in their respective markets that has encouraged the negative externalities that are the subject of this book. Only through the destabilization of their present domination in consumer attention, behavioral profiling, and filtered content dissemination can we break the hegemony that has forced these varied harms on the public—and effectuate market outcomes that would better protect the public interest. A number of options are available to the European and American regulatory community.

Natural Monopoly or Not?

In 2019, Facebook co-founder Chris Hughes[8] joined a chorus of voices (including presidential hopeful Senator Elizabeth Warren,[9] Senator Ted Cruz,[10] and former Secretary of Labor Robert Reich[11]) calling for regulatory agencies to apply antitrust authority to "break up" the digital giant. Hughes positioned the move as necessary to diminish the more damaging effects we have seen arise from Facebook and similar platforms, such as the spread of disinformation and hate speech. His argument drew a great deal of popular support from the broad public as well as sharp criticism from the competition policy establishment.[12] The company itself also pushed back, with newly minted public policy executive Nick Clegg penning a response arguing that breaking up Facebook would only serve to punish an innovative company that has created tremendous economic value.[13]

Such a "break-up" policy has taken center stage in the arena of national politics and could potentially involve splitting Google off into its constituent consumer internet parts, including Search, YouTube, Maps, Waze, and other media properties operated by Alphabet. Senator Warren, for instance, specifically proposed as part of her campaign that as president she would appoint regulators who are committed to using existing tools to unwind anticompetitive mergers, including Whole Foods and Zappos (Amazon); WhatsApp and Instagram (Facebook); and Waze, Nest, and DoubleClick (Google). Warren contends that "Unwinding these mergers will promote healthy competition in the market—which will put pressure on big tech companies to be more responsive to user concerns, including about privacy."[14]

Such break-up policies have been perceived by many as the most radical approach that could be considered when contending with internet firms' monopoly power. But is this really the most radical means of regulating internet firms and curbing their overextensions across markets?

I suggest an even more radical approach: Facebook and firms like it have become natural monopolies that necessitate a novel, stringent set of regulations to obstruct their capitalistic overreaches and protect the public against ingrained economic exploitation. While this option does not exclude the possibility of also pursuing a policy of breakup, I believe it is the more important objective and must take precedence. To understand why, we can apply rules of thumb from traditional competition and antitrust policy analysis, in which policymakers consider in a step-wise manner the economic dynamics of the industry in question.

Traditional American execution of competition policy involves first asking the question, Is this market competitive? Typically, the regulator might measure this by determining whether one firm controls a share of the market large enough to impose economic inefficiencies that unfairly favor the firm in question. In the United States, the Justice Department's Antitrust Division and the Federal Trade Commission (FTC) typically consider the relative market share of the firm in question. (Which agency actually makes the determination depends on the particular circumstance.) There is no explicit threshold for how much market share a company can have before it is considered a monopoly, although the FTC typically does not scrutinize a company for monopoly power if it occupies less than 50 percent of a market,[15] while the lowest-ever market share determined to result in monopoly power by the European Commission was 39.7 percent.[16] If the market appears competitive, you typically leave it alone; if it is not, then you move to the next phase of analysis.

In the case of the consumer internet, we know from a quick silo-by-silo examination that the market is not competitive. Google, for example, controls the internet search market in the United States. Additionally, in the view of many Google has abused its market power in that subsector, which would then necessitate a governmental policy intervention.

Let us briefly apply this analysis to Facebook as well. The company's industry can be hard to define because its holdings and technological features change so quickly, as do those of the overall sector. The same can be said of the other large consumer internet firms. Facebook's many platforms, including Messenger, WhatsApp, Instagram, and the big blue app, cover social media, photo sharing, and messaging, among other industries. I would contend that, in the United States, Facebook has a dominating monopoly presence in each of its principal consumer areas of activity. This in turn suggests that the primary markets in which Facebook operates are no longer competitive.

To move a competition policy inquiry forward in the United States from there, the company in question must not only be deemed to have excessive market share, it must also be shown to use its market position to exploit the consumer—a standard that the judicial system has over time taken to mean consumer price hikes. Many experts have contended that companies like Facebook and Google do not engage in such exploitation, because consumers are not charged money for access to even the highest revenue-generating services, such as social media and search—but this conclusion is squarely wrongheaded. Various forms of currency have lubricated various markets throughout history. The currency extracted from individuals in the consumer internet context is typically not money, but rather a novel, complex combination of individuals' personal data and attention (as discussed in chapter 3). Given the market concentration of leading consumer internet silos such as search and social media, one could suggest that Facebook and Google are pseudo-monopolistic firms that collect so much data and exploit so much of our attention that they systematically leave consumers in the lurch as a general matter of doing business. These firms are two-sided platforms that have monopolized the consumer side; accordingly, on one side of the platform they extract the end consumers' currency at extortionate monopoly rates, and on the other side of the platform they exchange it for monetary revenue at tremendously high margins. It is this subtle but corrosive form of exploitation that we should find most objectionable. Further, as many have argued,[17] there is a case to be made that we must not only consider the matter of extortionate rent collection as the threshold for regulatory intervention but also the perpetuation of subversive effects across the rest of society outside the corporate-consumer economic relationship itself, such as negative impacts on the labor markets, quality of services, and business innovation. Facebook and Google, among other internet firms, have likely caused such diminishments, necessitating a hard look at their effects on American market competitiveness.

Given a lack of competition in a market and established exploitation, the regulator's next question traditionally is, If the market were to feature many firms, could we imagine them all competing with each other in a dynamic, vibrant economic environment? Could fledgling entrepreneurs and other firms someday compete in the industry with the likes of the monopolists? If so, then a series of "pro-competition" policies should be pursued, perhaps ranging from issuing targeted regulations that narrowly protect consumers from certain forms of harm to actively breaking up the companies.

Conversely, is the reality that we can only ever imagine one player gaining power and subsequently maintaining that power until it reaches monopoly status, because of the fundamental nature of the market's interaction with society? This is essentially tantamount to asking, Is the firm a *natural* monopoly, or not? Railroads, roadways, and telecommunications firms have in the past been deemed natural monopolies because their costs of infrastructure and other barriers to entry suggest there should be no more than one player in each of those particular markets. In each case, one firm has emerged to control regional networks because it has not made sense for society—that is, the collective of American consumers in a given region—to invest in more than one telecommunication network or electric grid. Because why as a society would we want to build two such networks and have them run side by side? Instead it makes far more sense to collectively invest in only one and should that one firm engender perverse economic effects that harm society—perhaps through exploitative rent-seeking or intentional reductions in service quality (which American history has illustrated has been absolutely inevitable)—then we should regulate it and curb its damaging business practices such that economic gain is better distributed across society. It is this framework that engendered the phenomenon of public utilities—typically enterprises that feature some natural monopoly tendencies and are so critical to society's functioning that their operations must receive strict governmental oversight through independent regulatory bodies such as the Federal Energy Regulatory Commission and the Federal Communications Commission. When policymakers reach such a conclusion about a firm, they typically attempt to institute a set of rigorous "utility"-like regulations so that consumers are protected from exploitation before considering other ancillary forms of regulatory policy. The federal government applies such stringent standards to telecommunications firms, which must maintain minimum levels of service for emergency calling and provide guidelines for their handling of consumer data.

I believe that the consumer internet is a new kind of natural monopoly. Its leading constituent firms consistently exhibit network effects:[18] the networked services operated by Facebook, Amazon, and Google increase in value when more users use them. Meanwhile this makes it extraordinarily difficult for new entrants to offer competitive levels of utility to consumers out of the gate. As with the telecommunications industry before it, the consumer internet industry now maintains impossibly high barriers to entry. The leading internet companies have gradually established intricate, proprietary physical and digital infrastructures through the placement of new physical networks, the cultivation of preferential access to broadband providers and content owners, and the creation of an exclusive consumer "tracking-and-targeting" regime that necessarily shuts out the competition from access to the market for consumer data and attention. Furthermore, if a new entrepreneur does develop an innovative idea that picks up to a degree, an internet monopoly can readily acquire[19] or copy[20] it, as well as integrate it into its existing infrastructure.[21] Given this, there is very limited if any capacity for a second firm to effectively compete against consumer internet firms in the market silos they dominate.

If policymakers can eventually accept the economic circumstances that have now emerged and can agree that firms like Facebook and Google represent natural monopolies, we should then begin to consider utility regulations that can effectively hold them accountable to the public. In the past, the United States has given such designations to both private and public monopolies (including for instance electric utilities) that have variously resulted in the creation of new regulatory agencies to treat monopolistic overreach. In the case of consumer internet firms, such regulations could entail stricter standards concerning user privacy and data processing; clear and consistent investigations into any proposed merger, acquisition, or growth of business into parallel industries, especially in cases where excess concentration or market bottlenecking could result; complete transparency into the ways that the industry's algorithms disseminate ads and content, particularly to marginalized classes of the population; taxes or stipulations to uplift public interests such as independent journalism and digital literacy; and minimum-required investments into technologies that can detect and proactively act against obvious instances of hate speech and disinformation.

Implementing these regulations does not mean that we should not also pursue breaking up these firms. But doing so before considering what can be done to more directly curb their monopoly power may not effectively address with immediate effect the harms wrought by consumer internet firms—and as such, breakup is economically a lesser imperative.[22] What the U.S. government should in fact pursue are the overreaches of a business model that has systemically subverted the public interest and perpetuated a series of negative externalities in our media and information ecosystem.[23]

In the sections that follow, I present three targeted reforms that can

The Cyclical Nature of Natural Monopoly Development and Retention in the Consumer Internet Industry

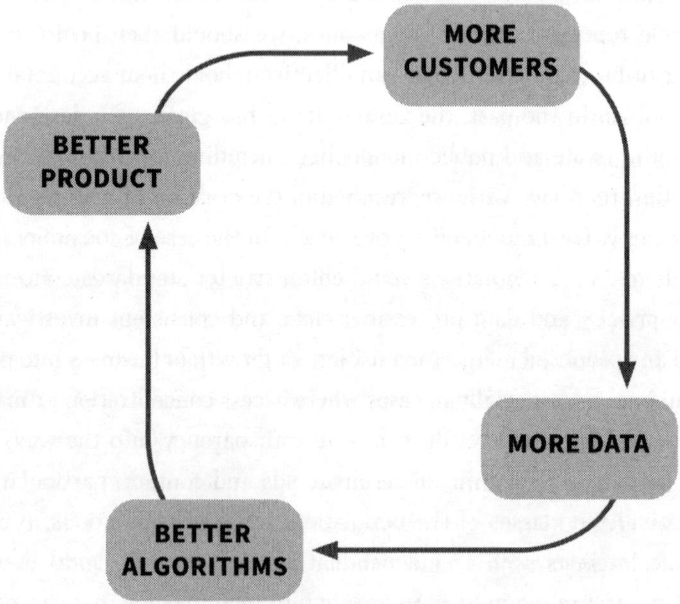

Source: Bjarke Staun-Olsen, "The Value Isn't Your Algorithm, It's Your Data," Creandum Medium, November 2016 (https://blog.creandum.com/the-value-isnt-your-algorithm-it-s-your-data-dda4a1bf688b).

help the regulatory community begin to address the harms to competition wrought by the internet sector.

Antitrust Enforcement and Reform

The present mode of antitrust enforcement—characterized by a strict and narrow focus on harm to consumers, particularly on price—is ill-equipped to assess the damages pushed by the digital-advertising giants onto consumers unless an earnest attempt is made to ascribe a value to the proprietary data that firms maintain on the individual. Even if this can be done, the possible second-order effects of a lack of competition in markets remain, in the form of negative outcomes on the labor markets and diminishments in product quality. American policymakers should buck the trend of the jurisprudence cemented in the federal courts and commit to giving the U.S. competition enforcers the regulatory bailiwick sufficient to examine and act decisively against the leveraging of monopoly.

This could be accomplished either by a much-needed reassessment of the Federal Trade Commission's 2015 "Statement of Enforcement Principles Regarding 'Unfair Methods of Competition Under Section 5 of the FTC Act,'" which outlines the agency's conservative approach and thus affords the courts full authority to maintain adherence to the narrow view of consumer harm, or through broader legislative reform efforts.[24] Such a reassessment can allow regulators to apply appropriate and effective remedies to combat anticompetitive practices, be they the mandate against certain harmful business practices, the imposition of critical breakups, or the requirement for a duty to deal. Furthermore, the Justice Department's Antitrust Division has broad authorities to enforce Section 2 of the Sherman Antitrust Act,[25] and while interpretation of the authorities afforded to the government by that law has over time narrowed in focus, there is no reason that the administration cannot attempt to buck the trend, which scholars argue has resulted in intense concentration in the digital economy. The presidential administration can work with the Antitrust Division to innovate its regulatory and legal approach into the future. Legislation is another possible avenue; a more direct route to overriding the conservative approach of the courts would be to work to pass a federal law revising the purpose and capacity for antitrust enforcement.

Scrutiny over Mergers and Acquisitions

It is clear that certain acquisitions have been permitted that perhaps should have received far greater regulatory scrutiny. Section 7 of the Clayton Antitrust Act and amendments to the act through the Hart-Scott-Rodino Antitrust Improvements Act were passed to give the U.S. government the authority to review mergers. Regrettably, they have done little to curb the acquisition of several key businesses in the digital sector, among them DoubleClick, Instagram, Waze, and WhatsApp. Some have claimed that even Amazon's acquisition of Whole Foods represents a significant problem in regard to the forward amalgamation of data in ways that could lead to consumer harm. Reports have also emerged that Facebook has aggressively gone about "squashing rivals, co-opting their best ideas or buying them outright as it cemented its dominance of social media."[26] The government should consider ways to improve the mechanisms for possible enforcement against such combinations that might eventually result in broad consumer harms. In particular, regulators should be fully equipped to understand and analyze the potential consumer harms associated with data sharing between merged entities. In addition to potential privacy harm to the individual, such combinations may also damage market competitiveness itself; data acquired through the purchase of a firm could, for example, be combined with the purchasing firm's data stores and used for competitive advantage in parallel or ancillary markets. In this context, vertical integrations of tracking and targeting should be of special concern to the regulatory community. Such integrations have systemically encouraged the disinformation problem and other negative externalities now prevalent online.

Citizen Data Portability

A final remedy to treat the national problem of market competition could be to unlock the industrial stranglehold on the single resource that monopolistic consumer internet firms find so vital to their positions in their respective two-sided markets: the individual's personal information. These firms fundamentally treat consumer information as a proprietary creation that is bound with their intellectual property and competitive proposition to the market. Even if firms could argue that collection of consumers' personal data was essential to their project of feeding it to a machine-learning algorithm to draw out users' behavioral profiles and once constituted a legitimate intellectual practice, such claims can no longer be made. In fact, artificial-intelligence platforms accessible for free off the shelf are capable of learning a behavioral profile, given the data. Furthermore, experts around the world have perhaps rightly pronounced that a customer's personal information—whether it is the raw data a customer has provided the platform through engagement and interaction or the algorithmic inferences the firm has made about the customer—should be that customer's property. A regime of earnest portability whereby the individual can move his or her data to a third party, be it another internet firm or a bank or healthcare provider, is a remedy the government should consider to induce the formation of a plurality of internet platform services.

Transparency

Experts contend that model designers can protect against bias through development of technologies that check the nature of the training data and fairness of the outcomes. But while such technological solutions can be engineered to counter the overreaches of learning models, what forces companies to be fair when it is in their commercial interests to discriminate, even unfairly so, as long as the discrimination is not illegal? I would thus suggest a different remedy: implementation of such technologies by the industry, backed up by accountability forced on the industry through smart and earnest governmental regulation.

Machine-learning technologies have come to the fore because of their tremendous efficiency. No longer do we need to monitor traffic systems to avoid areas of congestion: Uber, Waze, and Lyft can accomplish the task much more effectively on an algorithmic basis. No longer do we need news editors to determine what information should or should not go front and center before our individual attention: Facebook, Apple, and Twitter can infer who we are and what we want to see and route the relevant content to us directly. No longer do we need to ask the contracting expert what flooring suits our apartment the best: Amazon will find out for us and ensure speedy arrival. And no longer do we need to rely on the guidance counselor to help decide what college to apply to and ultimately attend: Google can address all of these concerns.

As machine-learning algorithms and artificial-intelligence systems become more ingrained in our daily lives and influence our behaviors throughout the day, we become increasingly dialogical with the machine that underpins the consumer internet. Society's observable actions and behaviors are actively feeding the decisions executed by the machine, and the corresponding commercially driven decisions, in turn, influence our actions and outlook in the real world. However, beyond the obvious questions this presents regarding individual autonomy, psychological dependence, mental health, and the broader concern of empowering a civilization-wide overdependence on machine technologies and implicit bias against sentient real-world expertise is the apparent reality: machines are discriminatory by design. Indeed, the more discriminatory they can be—the more incisive their predictions about individual behaviors and the collective outlooks of population classes—the more they add to the industry's pocketbook. This is Ad Targeting and Content Curation 101: if a machine can understand your mind, it is doing the job Facebook designed it to do. But in the course of so doing, the machine is bound to make frequent mistakes. There is no real-time learning system that can effectively model human psychology without making mistakes along the way—and it is in the noise that pervades the system where harmful bias lurks.

There is no reasonable solution, then, other than to use the full agency of the public interest to intervene and clarify for commercial entities what is right and what is wrong. In the absence of explicit rules and regulations, it is in the industry's best interest to breach the public interest so long as it is legal to do so and unintelligible to the public. If Instagram leaves such opportunities on the table, Snapchat will pounce, and vice versa. Sound business sense requires that both companies take up such opportunities unless the consumer market reproaches them through expression of collective sentiment in the marketplace or unless the government intervenes. And consumer outrage expressed through purchasing behaviors will take too long to be effective or will have minimal long-term impact in a space that offers little transparency. We need look no further than the voluntary reforms instituted by Facebook since the Cambridge Analytica revelations. While Facebook has ceased certain activities, firms that are not under the public eye have taken them up, taking advantage of commercial zones of operation cast aside by Facebook on the back of vociferous public advocacy.

We can observe the industry's actions in the face of governmental inquiries. If the industry earnestly wished to protect against these harms, why would it not wish to submit to governmental review and sectoral oversight? It is a case of the interests of private commerce versus the interests of human rights. The culture engendered by the Facebook maxim "Move fast and break things" signals machine bias and other challenges wrought by the radical capitalism seen in this industry. The consumer internet industry's tendency is to avoid challenges presented by algorithmic design until it becomes popular to attend to them—by which time it is too late. The industry's systems may have by that time contributed inordinately to systemic bias, prominent as it is in the American media and information universe.

A novel approach for governmental intervention should include four essential elements: federally funded research into ways to protect consumers from the harms of algorithmic bias; federally endorsed multistakeholder standards development toward a guiding framework for ethical artificial intelligence; industrial auditing and oversight of high-impact commercial internet algorithms, backed by governmental enforcement to ensure fairness; and safeguards to protect radical data and algorithmic transparency. The sections that follow discuss these proposed measures in increasing order of political difficulty, given the inevitable policy pushbacks each would face.

Transparency in Digital Political Advertising

Let us start from the same premise that economist John Kenneth Galbraith did concerning persuasive advertising: often consumers are encouraged to choose suboptimal options.[27] If internet companies cannot stop the flow of disinformation over their networks, the least they can do in the interest of democracy is to have the courage that Jack Dorsey of Twitter has shown and indefinitely block all political advertising until they can be assured that disinformation will not flow over their platforms.[28] Pinterest is another company that has blocked all political advertising because it believes it is not ready or qualified to handle or contain the potential spread of disinformation over its platform.[29] But other companies in the sector, including Facebook and Google, have not taken as progressive an action, opting instead to consider only disabling some microtargeting options for political advertising.[30] The two primary reasons behind this calculation are very clear: the firms do not wish to forego the ad revenues that come from the political sector, and they do not wish to upset any political camp with such an action—in this case, conservatives in the United States. (Twitter, for instance, felt backlash from the political right, including from Donald Trump's reelection campaign chairman Brad Parscale, in the immediate aftermath of Dorsey's announcement.) Indeed, a 2017 study by Daniel Kreiss and Shannon McGregor found as much through a series of interviews with political operatives and staffers at the technology firms. They reported that "technology firms are motivated to work in the political space for marketing, advertising revenue, and relationship-building in the service of lobbying efforts" and that "these firms have developed organizational structures and staffing patterns that accord with the partisan nature of American politics."[31] Firms understand well that, as legal scholar Ramsi Woodcock puts it, consumers buy products—or are convinced by persuasive political arguments disseminated by a campaign or disinformation operator—that are not the best option for them due to mass "psychological manipulation."[32]

But in the absence of such voluntary action on the part of the industry, government can step in to improve the citizen's situation. One of the most significant changes that the industry could pursue is greater transparency applied to digital advertising. In certain contexts, politics and elections among them, the purveyors of information deserve a degree of accountability for their decisions regarding the spread of information. Political communicators tap into social issues to appeal to the individual voter's psyche and charge the voter to make a certain decision at the ballot box. As they engage in this sort of activity, elections regulators are behooved to ensure that there is no nefarious intent behind these efforts.

A transparency regime will be critical in enabling the public—including journalists and academic researchers—to understand the historical extent and impact of disinformation operations. Such an effort should be focused on several key disclosures within the context of political advertising itself (not in a database on another website), including a visually prominent disclosure that the ad has been placed by a certain political actor; the disclosure of the source of the ad; the party responsible for funding the ad campaign; and measures concerning the reach and viewership of the ad campaign. Further, to enable research and analysis, firms should establish and maintain databases with API features.

That said, there exists no transparency regime that can universally protect the public because active disinformation operations engage in

Example of an Ideal Digital Ad that Clearly States the Ad Is Political and Includes a Specialized Disclaimer

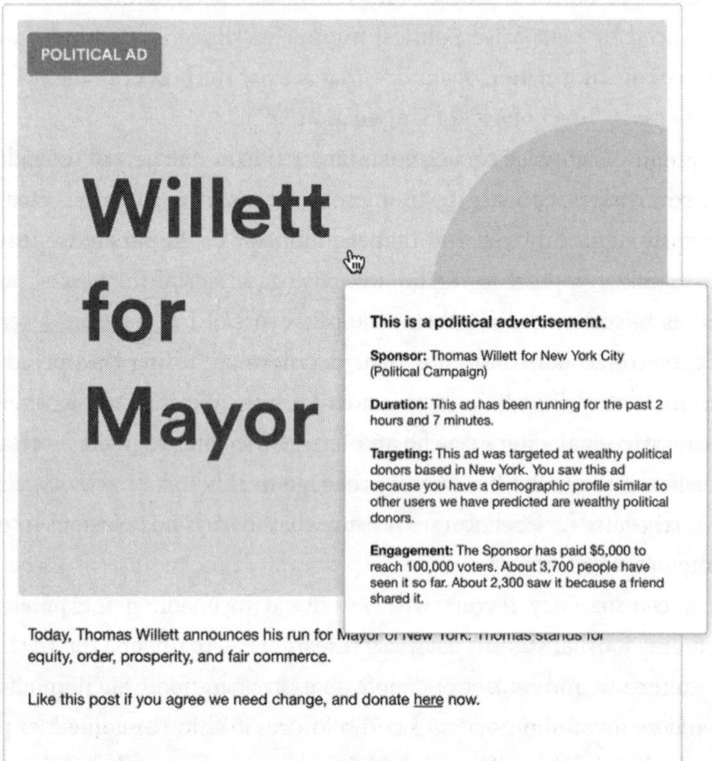

Source: Dipayan Ghosh and Ben Scott, *Digital Deceit II: A Policy Agenda to Fight Disinformation on the Internet*, September 2018 (www.newamerica.org/public-interest-technology/reports/digital-deceit-ii/competition/).

political deceit in the moment. Adopting better transparency in the internet industry is an indirect solution, as the impact it can have must be conveyed through the efforts of a small interested group of people who write and research about the impacts of political advertising. It is not a direct remedy but rather must be seen as an important and perhaps necessary piece of the puzzle of regulating the consumer internet.

New Ideas to Protect from Algorithmic Bias

In recent years, computer scientists have developed novel techniques to contend with machine bias.[33] But as a general matter these approaches variously cater only to certain types of models or are otherwise infeasible owing to companies' tendency not to question the fairness of their models before deploying them or because of other practical problems in the sector. More research is needed to develop industry-grade mechanisms to help protect against machine bias. We must also develop greater understanding regarding the impact of computing machines on society and what public policy measures should be taken to counter industrial overreach and contain harms. A good start has come from the Explainable Artificial Intelligence program, launched by the Defense Advanced Research Projects Agency. The government should channel further resources to such pursuits.[34]

In the way forward, government should more effectively support researchers in gaining access to corporate databases maintained by the internet firms. The industry's data-sharing programs with academic researchers have come to a relative standstill, and most data-sharing proposals fail to go through because the firms see minimal benefit to such partnerships. Research findings that support the commercial interests of the industry are unlikely given the research desires of most independent academics. Meanwhile, research findings could, as they typically do, implicate a firm's past decisions or current business practices. Furthermore, firms wish to avoid another Aleksandr Kogan–type incident, whereby the University of Cambridge researcher legitimately obtained private data from Facebook's platforms for 80 million users for apparent psychographic analysis and research purposes but then sold the data to the political advisory firm Cambridge Analytica, which may have used it in its engagements with clients, including the Trump campaign in 2016. The industry's hesitations are thus clear. Facebook's failure to meet its commitment to make available a dataset within a reasonable amount of time as part of its agreements in the Social Science One project is only one illustration of this.[35] My colleague Tom Wheeler has presented a more radical but fitting approach that generalizes the transparency use case from academics to the general public—an idea that features implementation of open, public-interest APIs that enable users to see what goes in and what comes out of social media algorithms.[36]

Policymakers must step forward and more formally support the research community in holding the industry accountable to share data with researchers. While the industry might argue that it is proprietary and private, designing and fairly administrating such sharing is the only measure that can give independent researchers the capacity to understand what impacts digital platforms are having on matters concerning our political discourse and societal circumstance.

A Design for Ethical Artificial Intelligence

In recent years mathematicians and computer scientists have variously come together with ethicists and philosophers from across the industry, civil society, and academia to produce a slew of ethical codes for artificial intelligence and machine learning. That these discussions exist is a positive development. But we must ensure that history does not regard them as fluff. One commonly cited framework highlights five key principles: responsibility, so that those with grievances regarding an algorithmic outcome have redress with a designated party; explainability, so that the algorithms and data used to develop them can easily be made clear to the public or those subject to their decisionmaking; accuracy, so that the model's errors can be identified and proactively addressed; auditability, so that third parties, including public interest agents, can investigate the algorithms and ensure their integrity; and fairness, so that the models do not perpetuate biased outcomes.[37] This represents a basis for a comprehensive set of principles on the governance and execution of fair machine-learning models.

The government should work with these and other stakeholders to coordinate a multistakeholder conversation concerning the development of an ethical framework for artificial intelligence. These conversations should be focused on the technological nature of the algorithms and data inputs themselves, leaving other important but less relevant contemporary conversations regarding the technology industry and the governance of artificial intelligence to the side. The government can use the National Institute of Standard and Technology's Cybersecurity Framework, developed under the auspices of the Obama administration, as a blueprint for such multistakeholder guidance.[38] Of particular importance throughout the process will be the guarantee that public-interest advocates are represented.[39] Such conversations can focus on the industrial use of artificial intelligence over the internet to maintain group focus while also addressing internet algorithms' outsize influence over the information ecosystem.

Oversight of Internet Algorithms

Consumer internet firms extensively implement machine-learning models to drive growth, engagement, behavior profiling, and revenue collection and management—among many other activities. These have a tremendous impact on public interests from fairness to democratic process and should be subject to general governmental oversight in some capacity. A model like the framework settled by the Federal Trade Commission with Facebook and Google through various consent orders may be appropriate. In response to industry overreaches, the agency settled new conditions with each company, including the ongoing auditing of their practices with regard to maintaining consumer privacy.[40]

This condition from the consent orders effectively enforced a profound change on the companies by compelling them to install what are now known as privacy-program teams—staff who are charged by the company to work with product managers and engineers to understand every single proposed product innovation, including the most minor of features, and to help the subject firm coordinate a cross-functional decision as to whether the proposed changes would threaten the user's privacy. The personnel in the privacy-program teams interact with external professional auditing consultants, who verify the integrity of the privacy practices of the subject firm, and develop periodic reports shared with the federal regulators that help affirm that the subject firm's privacy program is effectively protecting users from privacy overreaches.

It could be argued, in light of PricewaterhouseCoopers's failure to find Facebook's missteps that led to the Cambridge Analytica hacking, that these types of setups are bound to fail.[41] They can suffer, for example, from the traditional auditor's paradox, by which the auditing firm develops a close and collegial relationship with the subject firm and fails in its role as an independent review agency working in earnest for the public interest. Culturally, there can be a lack of incentive to report concerns accurately, largely because sharp criticisms will be seen by the subject firm. This is where the government can come in by holding all parties accountable. As the U.S. government pursues actions against the internet industry on the basis of further breaches of privacy, security, public trust, and algorithmic integrity, it should consider mechanisms to additionally force the companies to work with independent external auditors to ensure that internet-based artificial-intelligence systems are not putting public interests at risk.

Radical Data and Algorithmic Transparency

Centrally responsible for the excesses of the algorithms underlying the consumer internet architecture is the lack of transparency into how they are developed, how they operate, and what they accomplish.

Transparency into this regime—through consumer understanding of what data corporate actors hold on them, how behavioral profiles are developed through inference, how machine-learning models are used to develop such features as the news feed and YouTube's recommendation algorithms, and what the practical outcomes of these algorithms are—is critical to limiting the harmful discriminatory effects of the internet platforms.

Many have attempted to develop tools to layer such transparency over the sector—including the political ad–transparency projects led by ProPublica, Mozilla, and Who Targets Me, which were all discontinued by Facebook in early 2019.[42] That Facebook was so determined to block these services by tweaking its code illustrates the tension at the heart of true transparency measures: transparency attacks the business model of internet companies such as Facebook. These companies want to protect information on how their curation and targeting algorithms work for two primary reasons. First, allowing public access into the targeting metrics of Facebook ad campaigns would shine light on how the company's algorithms perpetuate bias of various kinds. Second, exposing the design behind algorithms enables the company's competitors to understand important strategic elements of Facebook's commercial makeup and adjust its strategies in real time to challenge the company's strength in the market. In other words, Facebook objects to transparency reforms as a form of commercial protectionism. I suggest a novel regime—a radical form of transparency I have presented with my colleague Ben Scott in a related work[43]—that can hold the industry accountable for the negative effects pushed by its models onto the public. Such transparency would enable users—or at the least, governmental or nonprofit organizations working in the public interest—to see the inputs and outcomes of the internet industry's algorithms.

The Bipartisan Case for Regulatory Reform of Internet Policy

A brief word on the partisan nature of regulation is necessary. We can acknowledge that increasing the regulation of industries has traditionally been a priority of Democratic politics, particularly in recent decades. The Obama administration and White House, through its prioritization of consumer-centered reforms in energy, education, technology, telecommunications, and healthcare, epitomized this.

Thus many of the regulatory proposals suggested in this chapter, as well as the tone and content of this book overall, might appear to favor one end of the political scale. I would challenge this by noting that the analysis I offer attempts to dissect the microeconomic situation in clear terms and offer a direct and earnest regulatory response to any evidence of exploitation. Theoretically, any entity could have been implicated—consumers, governments, Silicon Valley giants, start-ups, foreign nationals, and the foreign business community included. But the results of our analysis suggest that the internet industry centered in Silicon Valley is our target for reform.

Nevertheless, the broad strokes of political discussions concerning the technology and telecommunications industries in recent years have fallen roughly along party lines. Republicans like Federal Communications Commission Chairman Ajit Pai have supported the business interests of the industry, particularly internet service providers, by blocking the many attempts to push net neutrality policy over the past ten years.[44] They have supported the idea of free speech, pointing the finger at the major internet companies for alleged anticonservative bias revealed in a 2016 report about Facebook[45]—the implications of which Facebook chose to engage.[46] Meanwhile, since 2016, many Democrats have pushed the idea of increased regulation, with the Honest Ads Act perhaps being the most notable example of a commonsense regulatory proposal. Of course, there are many exceptions to these generalizations on both sides. Conservatives might suggest that the Obama administration was not as harsh on the tech industry as it could have been—suggestions that the Obama administration would most likely challenge. Furthermore, some in the Democratic Party may have taken a lenient stance toward the tech industry in anticipation of the 2020 elections and the desire to claim the industry's political donations. Meanwhile, as I have referenced numerous times, many Republicans have spoken out against the privacy and competition harms perpetrated by the tech industry.

Given such stiff partisanship in Washington in recent years, it might be the case that some will pick up the proposals suggested in this chapter and dispose of them out of hand. For those who are so inclined, I suggest the following: Consider the state of the constituent, the person who is consistently exploited, monopolistically hoodwinked, and unaware of the economic situation at hand. Consider that the constituent's economic currency—which comes in the aggregate complex form of the consumer's attention and data—is being raked at the hands of Silicon Valley's internet barons at fully hedonistic rates. How long will we continue to allow the economically powerful ranks to consume the world and the interests of the citizen? How much longer will we accept a toothless regulatory regime that fails the American consumer? We wish not to craft a spiteful regime that incites a cobra effect against the American economy at large. Likewise, we wish to avoid inducing a circumstance of differential technological development, whether over a geographic or, as suggested by Swedish philosopher Nick Bostrom, industrial gradient, through issuance of ill-conceived or underdeveloped regulations.[47] We want only to restore the fair economic and political functioning of the nation. It is time to earnestly come together and develop a fair proposal that puts American citizens at the heart of our political considerations.

It is all our fault that we have let things come to this. It is all our job to correct our error.

Industry Lobbying and the Logic of Collective Action

Years ago, Mancur Olson argued in his treatise *The Logic of Collective Action* that the interests of the well-organized few are typically over-represented in the halls of democracy while public interests—those interests that are diffuse across the civilization—feature poor political organization and therefore poor coordination in advocating policy changes.[48] As the scope of the interest in question grows larger—ultimately to the point that it can be construed as an interest of the public at large—the cause develops a free-rider problem: the more commonly the cause is shared among citizens, the more easily people can ride the coattails of those who actively advocate the common cause before policymakers. Thus as support for the cause grows across society, free riding intensifies—particularly since public goods are not excludable. But for those smaller groups that have a narrower collective interest that benefits only the members of the group, there will naturally be a greater return on advocating the group's cause. In such smaller groups, there will thus be diminished free-rider effects, as the few who are group members all stand to benefit. The small group can organize its cause quite efficiently and to great effect with little effort, but the big group—say, the public—will face a tremendously high cost in organizing its advocacy. In the end, this unfortunate behavioral circumstance of aggregate laze enables narrow causes to defeat public causes far more often than we might imagine.[49]

With regard to the topics discussed in this chapter, the consumer internet industry—led by Facebook, Google, and Amazon—is the narrow cause. The three firms and the broader industry benefit from the same high margins yet damaging business model. The industry's political position is clear: fight tooth and nail against any regulatory enforcement or legislative action that would force the companies to bend to the will of the people in the protection of privacy or market competition.[50] The opposition, meanwhile, is the general public, represented by its government, which has such diffuse interests as the maintenance of democracy, the safety of human life, and the assurance of fairness throughout society. If we apply Mancur Olson's theory to our situation today, we must conclude that we have been unable to enact reforms against the industry in large part because our interests are so varied, we are unevenly informed about the capitalistic overreaches of the industry, and many of us will free-ride on the efforts of those who are better informed and equipped to advocate for change in Washington regardless of the magnitude or nature of the public's political action. When the internet industry decides to throw a spanner in the work of potential reforms, it can mobilize legions of adroit interlopers to front its efforts to interfere with legislative action.

Notably, though, it does not have to be this way in the United States. To defeat the logic of collective action, or at least to tip the scale in favor of the public over the industry, will require only that our leaders—our politicians and business leaders—stand up as firebrands, like Senators Josh Hawley, Elizabeth Warren, and Mark Warner have done, and point out the overreaches of the consumer internet business model. We must stop letting the sleeping dogs lie and stand up for ourselves. Only as more and more Americans begin to see the truth can we build sufficient motivation to break the black box at the heart of Silicon Valley.

The Dilemma of Attending to Content Policy Reform

The world is focused on the matter of passing content policy reforms aimed at keeping offending content off internet firms' platforms. As clearly indicated by recent events in many parts of Europe, Sri Lanka, India, Brazil, and throughout the United States—most noteworthy, perhaps, the genocidal conduct of Myanmar military officials—internet firms need to do more to keep their platforms free of disinformation, hate speech, discriminatory content, and violence.

Open discussion of content policy regulations is a black hole of sorts in our present political climate. There are two reasons for this. The first is political conflict, particularly in the United States, over how and how much we should maintain the national commitment to free speech in the forum of consumer internet platforms. The debates surrounding First Amendment rights are rife with controversy in the United States, where conservatives have raised deep concerns over the account suppressions of such far-right thought leaders as Richard Spencer, Jared Taylor, and Laura Loomer, who have spread the spirit of white supremacy through their tweets, posts, and videos.

Second—and I believe more critically—the standards applied in regulating hate speech, disinformation, and other classes of offending content will vary greatly throughout the world. Different cultural norms, ranging from the fairly liberal to the ultraconservative, exist between countries and within countries. It will be a monumental challenge for civil society, national governments, and international organizations to arrive at a set of norms that the internet should apply in perpetuity. This is a task that will necessitate many hard discussions over many years—and even after all of that deliberation, we may have no clear path forward on the syndication of global standards.

I therefore contend that we should treat the matter of content policy as a separate issue entirely. We must categorize policy discussions around hate speech, disinformation, terrorism, and the like as matters of content policy and treat them independently of the second class of regulations that I have advocated in this book: economic regulations focused around privacy, transparency, and competition that attack the business models of Silicon Valley internet firms.

To be sure, both classes of future regulations—content policy and economic policy—are vital, and they are equally important. Society desires peace, but our internet platforms sow chaos by enabling the spread of disinformation, sow hatred by enabling the spread of white supremacists' messaging, and sow violence by giving authoritarian officials a platform to spread racist conspiracies. Society desires fairness, but our internet companies systematically exploit the individual, artificially and unjustly disable the vibrancy and dynamism of open markets, and make questionable decisions behind our backs.

But the public, politicians, and regulators the world over have focused primarily on content policy regulation. The reason is understandable: politics and public perception are focused on the here and now. At the same time, we cannot leave by the wayside the matter of economic regulations—policies that target the corrosive business model of the consumer internet firms. We cannot allow our consternation over the Russian disinformation campaign and the industry's ill-placed adjudications about what should or should not stay online to override the deeper concern at hand: it is the business model of the consumer internet that engendered and maintains these harms. We cannot let our deliberations over the content policy regulations to stay our hand for deeper-lying problems. To treat these problems and contain them at their source, we must not only lash at the leaves of the weed. We must poison its evil roots.

This is why I reflect relatively little here on matters of content policy. While it is important to address content to limit discrimination, protect elections, and save lives, these are largely all administrative concerns that will ultimately be determined at the discretion of regional politics and culture. It is not an intellectual debate; drawing the lines of content acceptability is a determination of the collective attitudes of users in a given locality. In the meantime, internet firms can hire content policy executives who are charged with exploring users' concerns and reflecting them into the platforms' governance.

The consumer internet firms have decided that determining what constitutes offending content is a responsibility that will have to eventually be graduated out of the industry. Consider, for instance, Zuckerberg's seemingly benevolent proclamation that he does not wish to be the arbiter of truth. Did he say this out of concern for humanity? The answer is likely no: he does not want to be the arbiter of truth because he does not want the weighty responsibility to rest on his and his company's shoulders. Why should he take the blame for the Russians' activity on his platforms and its impact on the 2016 U.S. presidential election when we as a society cannot even determine what kinds of

The Logic of Internet-Based Content Negative Externalities

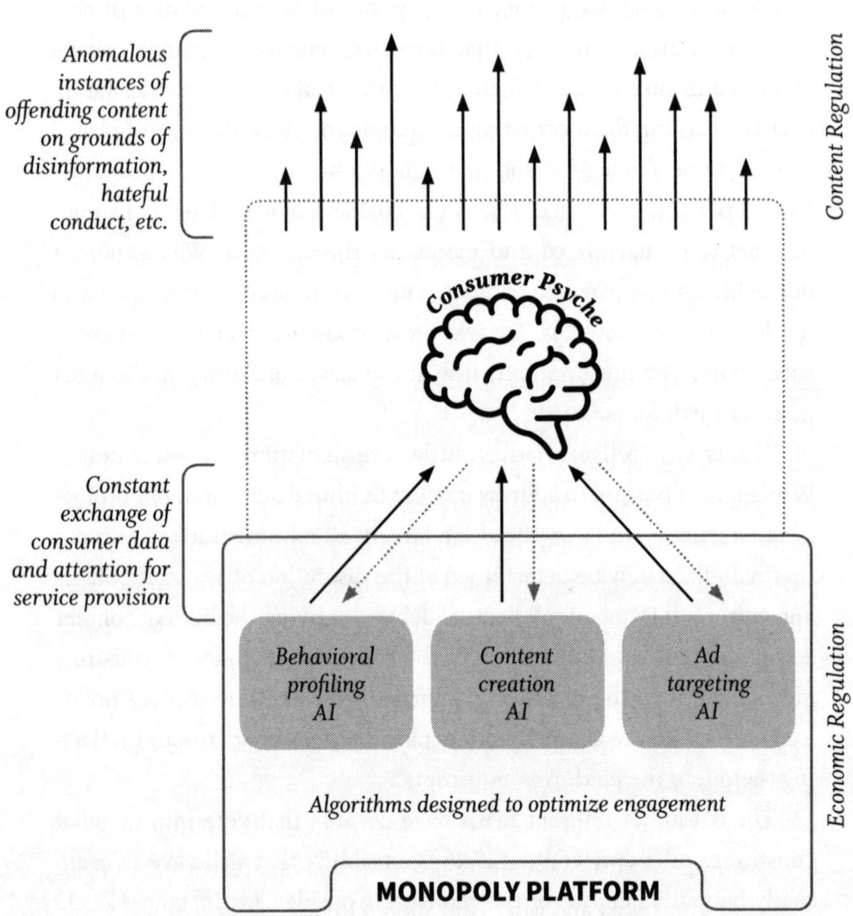

The business practices that underlie the digital platform encourage negative externalities, such as the disinformation problem; they are the root cause. As such, we require economic regulation—not content regulation—to limit the threats of these negative externalities in the long run.

content should be considered fake news? Whatever the negative externality, he wishes to pass on the responsibility of making such determinations.

But pass on to whom? That does not seem important to industry executives, so long as it is a third party—an entity external to the firm—that has the public's trust. That third party could be a governmental agency, a civil society organization, an industry consortium, a review committee, or a nonprofit set up exclusively to resolve questions concerning offensive content. The organization should, in the industry's view, simply have authority and the public's trust in its local jurisdiction. It should be seen as the source of truth by the users of the platforms.

Herein lies one final concern. The industry knows that it will take an eternity to develop the many questions that such arrangements to address content policy challenges would necessarily raise—in terms of who should have such authority over content policy, how involved the regional and national governments should be in the decisionmaking processes, how to prevent political influence, and, perhaps most critically, just where to draw the line. Consider the situation in the United States, where Democrats and Republicans cannot even resolve to pass the commonsense policy advocated in the Honest Ads Act, which simply proposes imposing transparency over the provenance and dissemination of digital political ads. If we cannot find resolution on that issue after four years of deliberation, we are unlikely anytime soon to be able to develop the content policy standards that Twitter, Google, Snapchat, Facebook, and Microsoft should follow.

The industry knows this well. Its leaders are aware that heated debates around content policy will persist for a very long time given our political circumstances, and that while they persist, we will be less focused on the more fundamental problem of economic regulation. Their greatest fear is economic regulation. They fear true privacy, competition, and transparency standards that would force changes to their business practices, because such regulations, if earnestly designed to curb the exploitation of consumers, would seriously cut into their business models. This would jeopardize both their personal wealth and their shareholders' interests; any curbing of the business model would significantly diminish the firms' profit margins. How significant that margin reduction would be would depend only on how serious are the regulatory standards advanced. Consumer internet executives will secretly encourage public debates over fake news and hate speech. They will add fuel to these flaring deliberations for as long as they can, drawing our eyes away from the subtler, more fundamental problems at the heart of the industry's commercial regime. Facebook's proposed new "oversight board" is the perfect example of this. The board should be designed by the company to tackle not only questions of content policy violation but also economic overreach by the company itself.[51] Therein lies society's true demons.

All this is to say that we must always prioritize the question of economic regulations. Let the war room for strategizing the passage of comprehensive privacy law be our *point d'appui*. Let us not die on the battlefield of content policy regulation, which will entail a series of global debates that are unlikely to ever have a clear unifying international norm given varying political views, even within countries like the United States. For the more our attention is diverted to the problem of content policy, the less we will focus on curing society of the virus that lurks beneath—and the more its malevolence will spread.

Driving toward Global Norms in the Face of Rising Foreign Uncertainty

The call to regulate U.S. technology and internet firms—including Facebook, Google, and Amazon but also Twitter, Snapchat, Microsoft, and others—has raised inevitable questions about whether Western regulation will apply to foreign firms that serve Western audiences. Chinese companies TikTok and DailyMotion are included in this list, as are network and device manufacturers such as Huawei, HTC, and others. The concern is clear: how will the American technology industry remain competitive—particularly overseas—if it has to comply with economic reforms while Chinese firms do not? The Nikon choir that captured Mark Zuckerberg's unused argument from his notes during his Senate hearing in 2018 explicitly encapsulates the general concern: the U.S. government should not break up Facebook because "U.S. tech companies [are] a key asset for America," and break up only "strengthens Chinese companies."[52] Other concerns include the establishment of an unlevel playing field that disfavors American firms with respect to their foreign counterparts, which could result in uncompetitive market positions for American firms; possible diminishments to the long-term dynamism of the American media ecosystem; a split in global technological governance mechanisms, which could lead to an internet that effectively separates some parts of the globe from others; and national security risks that could be raised by the extension of services from foreign internet firms that have the help of adversarial governments to achieve market power in local foreign markets, which could offer those adversarial governments novel back doors into global information networks.

Indeed, these are the very concerns raised by the consumer internet industry: should Western democracies levy stringent regulation against the Western companies leading the technology sector, economic and national security risks will arise as Chinese firms extend their dominance to other nations. To Zuckerberg's general point, not in a month of Sundays should policymakers assume that Chinese firms will comply with stringent regulations enforcing competition, privacy, or transparency promulgated by Western democracies.

Meanwhile, American firms are also not clear on what stance they should take in the face of an oppressive Chinese government. China's recent launch of an invasive surveillance regime on foreign firms that wish to engage in business operations in China—a regulation that Apple has quietly complied with—is illustrative of the implicit security risk of doing business with China and the inevitable ethical quandaries

American firms will find themselves in. Indeed, Facebook and Google have seriously considered entries to China in recent years, too, and have likely only pulled out for the time being because of intense public pressure, which they face more so than Apple since entry to China for them would entail the installment of rigorous content censorship and governmental surveillance regimes that would not sit well with Western users and governments.

But the potential opportunity for a company like Google to launch a censored search engine in China that complies with governmental requirements—called "Project Dragonfly" within the company—is, in a word, monstrous. Let us assume that a current Google user is on average worth $47 annually to the company, which is equivalent to the company's $111 billion in total annual revenues multiplied by its ad revenue share of 84 percent and then divided by the company's two billion users. Further, let us conservatively assume that a potential typical Chinese Google Search user is worth $16 annually, which is one-third the value of a typical Google user. We can also assume that the total internet search market in China comprises 800 million monthly active users, which is growing at an annual rate of 10 percent and will reach up to 1.3 billion in five years, at which point it will plateau. Let us also estimate that Google will take a 10 percent market share per year after entry up to 50 percent, which then plateaus due to competition from Chinese firms. Finally, let us say that all costs including tax are equal to 70 percent of revenues. The total value of the project based on these estimates is then $38 billion, assuming a perpetual market and a discount rate of 70 percent.

We require no further information to understand the heated discussions that are probably happening inside the boardroom at Google. The tension is clear: tackle the gargantuan commercial opportunity at hand, or accede to the democratic-minded protests of the more traditional Western customer base? Zooming out from the specific case of Google's possible entry into China and extrapolating the current global economic circumstances into the next decades, such tensions are the precursors of an inextricable Collingridge dilemma of cosmic proportions.[53] It is an international economic tradeoff that for the time being will continue to trouble corporate executives in the industry as it looms suspended over a flummoxed Silicon Valley.

It is perhaps anyone's guess as to which way the rational internet firm should bet; there are no clear answers to the questions raised here. But for the American government, projection of a clear and coherent plan outlining how we can maintain economic security for the American technology industry while also holding it accountable will serve as a counterforce to the industry's resistance. I suggest enacting the following policy actions, which can largely be taken independently of the new social contract, to protect the interest of American innovation while encouraging the international community to adopt the open governance model of the internet that is so critical to its functionality in service of democracy.

Make a Global Call for Regulatory Reform. As is now clear to the American public, regulatory reform is needed for Silicon Valley's breaches against the national interest. The administration should acknowledge this need, affirming the need for policy reforms aimed at transparency, privacy, and competition and stressing the importance of working with Congress to achieve them. Advocating such a far-reaching progressive new regime for internet regulation will send a strong signal to international allies that the United States is willing to hold even the strongest members of the American industry accountable—a position that should enable us to better partner with the international community in undertaking discussions concerning global economic reforms in presently contentious matters of technology, trade, and beyond.

Form an Internet Governance Coalition That Commits to Principles of Openness. The internet is splintering: Western and developed economies advocate an open form of governance; China and Russia take a more closed approach involving extreme forms of governmental censorship; and other developing nations practice something in between the two. The Obama administration's difficulties in negotiating the Commerce Department's management of the Internet Corporation for Assigned Names and Numbers is illustrative of the difficult policy issues involved. It is in the American interest to encourage countries that are uncommitted to adopt the American approach so that their information ecosystems are not absorbed by China and Russia. The United States should do everything in its power to develop and convene a U.S.-led coalition that would bring together the countries that practice open governance and apply pressure and motivations for uncommitted countries to agree to open governance standards. Washington should bring together the developed, open-governance nations, including Western Europe, Japan, Korea, Brazil, Argentina, and Israel, to set the tone, commit to codified governance standards, and discuss schemes to hold developing countries to the same standards.

Engage Developing Nations Intellectually on Internet Policy Reform. Developing nations such as India and Nigeria are undecided on whether to take a more closed or open approach to internet governance. The United States should commit to working with developing nations through the UN's International Telecommunication Union. In particular, it should build on the efforts of Phil Verveer and Danny Sepulveda, previous U.S. ambassadors to the agency who have engaged developing nations in robust discussions to encourage them to adopt open governance norms. The administration should also conduct a thorough analysis of the trade policy commitments it has made with developing countries and identify opportunities for more favorable trade deals or general circumstances in exchange for public commitments from such countries that they will adopt the open governance standards practiced by the United States.

Criticize the Approaches of China and Russia. In addition to the above, the United States should develop statements of administration policy that explicitly admonish the patriarchal internet governance standards practiced by the governments of such nations as China and Russia. Furthermore, Washington should use the tools it has at hand to sanction these approaches. Specifically, the United States should suggest a full review of any commercial technology dealings with either nation by the Treasury Department's Committee on Foreign Investment in the United States, which has recently released stringent policies against certain types of technology-related business engagements from these two countries. Additionally, Washington should continue to maintain America's rigorous abstention from Chinese and Russian technology and encourage other developed and developing nations to make similar commitments to shun networking equipment for telecommunications infrastructures from those countries.

Integrate Internet Governance Concerns into Sanctions Policy and Diplomatic Engagements. The United States should enforce these proposals by designing and implementing a novel sanctions framework that would effectively apply economic restrictions on nations that continue to engage closed governance standards and use Chinese and Russian networking technology despite the open coalition's guidance otherwise. Furthermore, the United States should commit to training its diplomatic staff on these matters of internet governance so that junior and senior diplomats can advocate the U.S. approach to internet governance and push back on foreign considerations to move in other directions. To accomplish this, Washington should build on the program initiated by the Obama administration to develop and dispatch diplomatic staff trained specifically on the digital economy to embassies worldwide.

Sponsor and Source Academic Research on the Importance of an Open Internet. The internet is central to the information ecosystem and defines the shape and trajectory of public opinion. Accordingly, most experts agree that the closed internet governance methods pursued by the Chinese and Russian governments is damaging for the local environment. That being said, there is very limited open, independent academic research on these issues that establishes that the U.S. approach to internet governance maximizes social welfare. The United States should thus commit to supporting more academic research in the areas of internet governance, transnational coalition building, and digital regulation to empower the academic field study of these issues. If such research reveals that open governance mechanisms are ideal, the administration should use it in its discussions with countries that remain on the fence in deciding on an approach to internet governance.

The American government must work with a broad domestic and international community to formulate and execute these policies. An open internet is generally seen as a favorable circumstance for U.S. industries, including the technology community. We should leverage Silicon Valley's full weight in supporting this message, particularly internet firms, including Facebook, Google, and Twitter, which are known to strongly support open internet standards. We should also bring in the full weight of civil advocates, a number of whom support such an approach to internet governance explicitly. With these two extremes—the internet industry and civil advocates—we expect to be able to organize a wide-based coalition to support the forward-looking policy positions and measures.

The battle over internet governance will be long and drawn out; an absolute win cannot be achieved in the near term because we can continue to expect emergent authoritative regimes to claim ownership and oversight of local communications networks. The best the United States can presently do is to use all of the policy tools we possess to encourage those nations that are unsettled in their approach to internet governance to adopt open norms.

The Case for Radical Reform

We have only entered the show; we stand long years away from its eventual denouement. The question we as a society must answer in the coming years is what form we wish that final act to take, and which direction to point our inquiries of the industry to accomplish the reforms we desire. American technology firms have fundamentally reshaped the media and information ecosystem, and in practice they determine and deliver the content that Americans consume and engage with. While tremendous economic benefits have accrued to the individual consumer and the business community alike, they have come at too high a cost. The calamities facilitated by the workings of the industry's digital platforms have struck with baleful force in recent years as the disinformation problem, the spread of hateful speech online, the systemic reach of algorithmic discrimination, and various other harms perpetuated by the industry's business practices. This has prompted a call for action on the part of the U.S. government—a call that originates from the indignant American consumer.

In this analysis, I have suggested a novel regulatory framework to respond to the capitalistic overreaches of the internet industry, including new regimes that would institute into the provenance of digital advertising such rights as transparency for the individual consumer; privacy protections from unfair data collection, processing, and use; as well as improved vibrancy and dynamism in the market through a slate of reforms promoting competition.

Some have contended that reform agendas targeting big tech will have been thrown off the train tracks by the coronavirus outbreak. These sentiments are driven largely by two primary factors. The first is that public attention has been absolutely absorbed by the outbreak. The second is that we have become more dependent on internet technologies through the course of the pandemic because of our need to stay connected to others for work and personal reasons, which could translate into a more favorable view of the internet sector. Some of the sector's activities, such as its data analysis and sharing programs, have encouraged this positive feeling about the industry's efforts.[1] Eric Schmidt has gone so far as to suggest that the virus will make big tech still bigger, noting that the "strongest brands and the strongest companies will recover more quickly," and that "the industry leader, if it's well managed, tends to emerge stronger a year later."[2] Google's announcement in April 2020 of its plan with Apple to enable app developers to use Bluetooth technology in smartphones to allow the owner to track every other person he or she comes into contact with is indicative of the kind of corporate strength Schmidt has alluded to. A cynical reading of these circumstances might suggest that the fact that firms like Apple and Google present the only options to lead contact-tracing efforts is just one example of the problems instigated by excessively concentrated market power in the technology sector.

These suggestions are true in the near term but mistaken in the long run. The significant privacy concerns that have been raised about the use of technology to monitor the outbreak suggests that we will need a comprehensive digital rights framework in the long run.[3] And beyond such concerns over surveillance coordinated through the government, we know that the outbreak and its aftermath will pass. It has without a doubt been disruptive, to the global economy and public policy and political action. But we will move past this virus and its effects, along with any temporary sentiments we might harbor regarding the industry. Sometime after the shock of COVID-19, life will return to normal, and with it, sentiments about the powerful position of the technology industry in our society. At that juncture, we will still be living in an age in which the modes of mediated communication are dominated by select internet platforms. We will once again see the economic reality before our eyes, and the damage that is being done to each of us with every passing day.

There is a global industry now behind the commercialized internet—and there are norms we believe in that will not foreseeably be adopted by our international competition. We must consider how to balance the design of a new and much-needed regulatory regime with the challenge of ensuring the competitive edge of American commerce, including both existing and future businesses. The one outcome we would wish to avoid involves the firms going Galt in response to the promulgation of American regulations that disengage them from global competition in the face of competing foreign internet firms that are not subject to such rules and regulations. Should our regulatory frameworks be ill-conceived, we will risk the departure of internet firms to greener pastures or perhaps even worse—their submission to foreign rivals.

Comprehensive privacy reform, instituting transparency into the collection and use of personal information, and a robust new competition policy regime, when undertaken as national policy and enforced against the industry, will indubitably redistribute the allocation of power between Silicon Valley, the government, and the individual. This would not be the first time a market trajectory has been redirected by government—nor will it be the last. But the harms wrought by the largest firms controlling the digital advertising space have become clear and are in public view; only progressive regulatory reform and the enforcer's stick can begin to counter such commercial and anticompetitive behaviors. What is most of all necessary in this new age of technology is the formulation of a new digital social contract that once and for all honors the rights of the individual.

The economic design of the United States rightly gives the capitalistic market free sway to innovate. But when such commercialization breeds exploitation of the individual, our nation has always acted to protect our democratic interests ahead of the freedom of markets. The preservation of our democratic process to favor the individual citizen's interest remains the country's core interest and the one unifying force behind the American national interest. Silicon Valley's internet industry must be treated no differently.

Notes

Introduction

1. "Premji Meets Buddhadeb over Wipro's New Campus," *Economic Times*, December 8, 2010.

2. Joshua Cooper Ramo, *The Seventh Sense: Power, Fortune, and Survival in the Age of Networks* (New York: Little, Brown & Company, 2016).

3. John Stuart Mill, *Principles of Political Economy with Some of Their Applications to Social Philosophy*, 7th Edition, edited by William James Ashley (London: Longmans, Green and Co., 1909).

4. Stephanie Nebehay, "U.N. Calls for Myanmar Generals to Be Tried for Genocide, Blames Facebook for Incitement," *Reuters*, Aug 27, 2019.

5. Mark S. Langevin and Edmund Ruge, "Bolsonaro's Base: The Social Liberal Party's Strengths in Rio De Janeiro Could Yet Become Weaknesses," *London School of Economics and Political Science*, Mar 13, 2019.

6. "Pablo Ochoa, Jair Bolsonaro: Why Brazilian Women Are Saying #Nothim," *BBC News*, Sep 21, 2018.

7. Anthony Faiola and Anna Jean Kaiser, "Rio's Pentecostal Mayor Takes on the Capital of Carnival," *Washington Post*, Dec 13, 2017.

8. Bryan Bowman, "White House Shrugs off Rise of Brazil's Far-Right Presidential Candidate," *The Globe Post*, Oct 17, 2018.

9. David Biller and Bruce Douglas, "Bolsonaro's Words Are the Sparks as Brazil's Farmers Burn Amazonia," *Bloomberg*, Sep 6, 2019.

10. Manoel Horta Ribeiro, "Auditing Radicalization Pathways on Youtube," *arXiv*, Sep 3, 2019.

11. Max Fisher and Amanda Taub, "How YouTube Radicalized Brazil," *New York Times*, Aug 11, 2019.

12. Vir Sanghvi, "India's Lynching App: Who Is Using WhatsApp as a Murder Weapon?" *South China Morning Post*, July 9, 2018.

13. Kunal Purohit, Misinformation, "Fake News Spark India Coronavirus Fears," *Al Jazeera*, Mar 9, 2020.

14. Gopal Sathe, "How the BJP Automated Political Propaganda on WhatsApp," *Huffpost*, Apr 17, 2019.

15. Julie Bosman, "How the Collapse of Local News Is Causing a 'National Crisis,'" *New York Times*, Nov 20, 2019.

16. Judy Woodruff, "How the Decline of Newspapers Creates 'News Deserts' around the Country," PBS Newshour, Jan 31, 2019.

17. Jon Allsop, "Another Brutal Week for American Journalism," *Columbia Journalism Review*, Jul 2, 2019.

18. Joseph Schumpeter, *Capitalism, Socialism and Democracy* (New York: HarperCollins, 1942).

19. Casey Newton, "New Legislation Is Putting Social Networks in the Crosshairs," The Verge, Aug 1, 2019.

20. Chaim Gartenberg, "Google News Initiative Announced to Fight Fake News and Support Journalism," *The Verge*, Mar 20, 2018; Mathew Ingram, "Facebook Says It Plans to Put $300m into Journalism Projects," *Columbia Journalism Review*, Jan 15, 2019; Lucia Moses, "One Year in, Facebook Journalism Project Gets Mixed Reviews from Publishers," *Digiday*, Jan 6, 2018; Mark Epstein, "Google and Facebook Worsen Media," *Wall Street Journal*, Feb 10, 2019.

21. "Christchurch Attack: Brenton Tarrant Pleads Not Guilty to All Charges," *BBC News*, Jun 14, 2019.

22. Matthew Schwartz, "Facebook Admits Mosque Shooting Video Was Viewed at Least 4,000 Times," *NPR*, Mar 19, 2019.

23. "Macron, Ardern Host Paris Summit against Online Extremism," *France 24*, May 15, 2019.

24. Henry Cooke, "Who Is and Isn't Coming to Jacinda Ardern's Paris Summit on Social Media," *Stuff*, May 12, 2019.

25. "Sheryl Sandberg Sits Down with CNBC's Julia Boorstin Today," *CNBC*, Mar 22, 2018.

26. "Facebook's Old Motto Was 'Move Fast and Break Things,'" *Mind Matters News*, Oct 19, 2018.

27. "Twitter Safety, Information Operations Directed at Hong Kong," *Twitter*, Aug 19, 2019.

28. "What Is Psychographics? Understanding the 'Dark Arts' of Marketing that Brought Down Cambridge Analytica," *CB Insights*, Jun 7, 2018.

29. Angela Chen and Alessandra Potenza, "Cambridge Analytica's Facebook Data Abuse Shouldn't Get Credit for Trump," *The Verge*, Mar 20, 2018.

30. Avery Hartmans, "It's Impossible to Know Exactly What Data Cambridge Analytica Scraped from Facebook—But Here's the Kind of Information Apps Could Access in 2014," *Business Insider*, Mar 22, 2018.

31. "Monopoly by the Numbers," *Open Markets Institute*, n.d.

32. Brian Winston, *Misunderstanding Media* (Harvard University Press, 1986).

33. Kaveh Waddell, "Facebook Approved Ads with Coronavirus Misinformation," *Consumer Reports*, April 7, 2020.

34. Mikael Thalen, "This Far-Right Troll Says Joe Biden Has Coronavirus and Will Die within 30 Days," The Daily Dot, March 23, 2020.

35. Will Sommer, "Twitter Suspends Conservative Huckster Jack Burkman over Coronavirus Disinfo," The Daily Beast, March 19, 2020.

36. "Misinformation around the Coronavirus," National Public Radio, March 14, 2020.

37. Hadas Gold, "WhatsApp Tightens Limits on Message Forwarding to Counter Coronavirus Misinformation," CNN Business, April 7, 2020.

Chapter 1

1. Brian Barth, "Big Tech's Big Defector." *The New Yorker*, Nov 25, 2019.

2. Shoshana Zuboff, "You Are Now Remotely Controlled," *New York Times*, Jan 24, 2020.

3. Ben Hoyle, "Tech Is Tearing Us Apart," *The Weekend Australian*, Jan 17, 2020.

4. Zeynep Tufekci, "Zuckerberg's So-Called Shift toward Privacy," *New York Times*, Mar 7, 2019; Jason Linkins, "The Death of the Good Internet Was an Inside Job," *The New Republic*, Dec 31, 2019.

5. Steve Lohr, "The Week in Tech: How Is Antitrust Enforcement Changing?" *New York Times*, Dec 22, 2019.

6. Lesley Stahl, "Aleksandr Kogan: The Link between Cambridge Analytica and Facebook," *CBS News*, Sep 2, 2018.

7. Julia Carrie Wong, "Former Facebook Executive: Social Media Is Ripping Society Apart," *The Guardian*, Dec 12, 2017.

8. Ben Gilbert, "Former Facebook Exec: I Take Back What I Said about Facebook 'Destroying How Society Works'—Kinda," *Business Insider*, Dec 15, 2017.

9. Timothy Lee, "YouTube Should Stop Recommending Garbage Videos to Users," *Ars Technica*, Aug 12, 2019.

10. Cat Zakrzewski, "The Technology 202: Googlers Demand Company Renounce Working with Trump Immigration Agencies," *Washington Post*, Aug 15, 2019.

11. Ali Breland, "Trump: Google, Facebook, Amazon May Be in a 'Very Antitrust Situation,'" *The Hill*, Aug 30, 2018.

12. Donald J. Trump (@Realdonaldtrump) on Twitter, Aug 28, 2018, 11:02 A.M.

13. Tony Romm, "Trump Again Accuses Social Media Companies of Censoring Conservatives," *Washington Post*, Mar 19, 2019.

14. Donald J. Trump (@Realdonaldtrump) on Twitter, Aug 19, 2019, 11:52 A.M.

15. Alina Selyukh, "Sessions to Meet with State Attorneys General about Social Media," *NPR*, Sep 25, 2018.

16. Nick Statt, "Google Could Face Far-Reaching Antitrust Investigation as soon as Next Week," The Verge, Sep 3, 2019.

17. Kate Cox, "The FTC Is Investigating Facebook. Again," Ars Technica, Jul 25, 2019.

18. Steve Lohr, "House Antitrust Panel Seeks Documents from 4 Big Tech Firms," *New York Times*, Sep 13, 2019.

19. Lauren Hirsch and Lauren Feiner, "States' Massive Google Antitrust Probe Will Expand into Search and Android Businesses," *CNBC*, Nov 14, 2019.

20. Nash Jenkins, "The Mark Zuckerberg vs. Ted Cruz Showdown Was the Most Explosive Part of Today's Facebook Testimony," *Time*, Apr 10, 2018.

21. Alex Ward, "Watch Sen. Graham Grill Mark Zuckerberg on Whether Facebook Is a Monopoly," Vox, Apr 10, 2018.

22. Maggie Miller, "Bipartisan Group of Senators Seeks to Increase Transparency of Online Political Ads," *The Hill*, May 8, 2019.

23. David McCabe, "Scoop: 20 Ways Democrats Could Crack Down on Big Tech," *Axios*, July 30, 2018.

24. Issie Lapowsky, "Elizabeth Warren Fires a Warning Shot at Big Tech," *Wired*, Mar 8, 2019.

25. "Democratic Rep. David Cicilline Discusses Probe into Big Tech Antitrust Regulations," *NPR*, June 6, 2019.

26. Todd Haselton, "Senator Says Facebook's Mark Zuckerberg Should Face 'Possibility of a Prison Term,'" CNBC, Sep 3, 2019.

27. Adam Satariano, "G.D.P.R., a New Privacy Law, Makes Europe World's Leading Tech Watchdog," *New York Times*, May 24, 2018.

28. Eric Lutz, "Digital Gangsters": British Lawmakers Torch Facebook, Accuse Zuckerberg of 'Contempt,'" *Vanity Fair*, Feb 18, 2019.

29. Graham Kates, "Global Lawmakers Investigating Election Meddling to Invite Zuckerberg to Hearing in Canada," CBS News, Feb 7, 2019.

30. "Competition Law Crosses the Digital Threshold," United Nations Conference on Trade and Development, July 18, 2019.

31. "The CNIL's Restricted Committee Imposes a Financial Penalty of 50 Million Euros against Google LLC," Commission nationale de l'informatique et des libertés, Jan 21, 2019.

32. Rob Nicholls and Katharine Kemp, "Consumer Watchdog Calls for New Measures to Combat Facebook and Google's Digital Dominance," The Conversation, July 26, 2019.

33. "Japan Government Begins Talks on New Legislation to Tighten Regulation of Its Giants," *The Japan Times*, Oct 4, 2019.

34. Leika Kihara, "Japan Sets Up Working Group on Impact of Facebook's Libra Ahead of G7," Reuters, July 12, 2019.

35. See, for example, Jon Russell, "Singapore Passes Controversial 'Fake News' Law which Critics Fear Will Stifle Free Speech," TechCrunch, May 9, 2019; and "Germany Starts Enforcing Hate Speech Law," *BBC News*, Jan 1, 2018.

36. "Facebook Challenges Belgian Tracking Ban," BBC News, Mar 28, 2019.

37. Alex Hern, "Italian Regulator Fines Facebook £8.9m for Misleading Users," *The Guardian*, Dec 7, 2018; John Oates, "Italian Data Protector Makes Facebook an Offer It Might Want to Refuse: A €1m Fine for Cambridge Analytica Data Leak," *The Register*, June 28, 2019.

38. Chris Merriman, "Google Faces €15m Fine for Netherlands Privacy Violations," *The Inquirer*, Dec 16, 2014.

39. "Brazil's New General Data Privacy Law Follows GDPR Provisions," *Inside Privacy* (blog), Covington & Burling, LLP, Aug 20, 2018; Diego Fernandez, "Argentina's New Bill on Personal Data Protection," International Association of Privacy Professionals, Oct 2, 2018.

40. Nicholas Confessore, "The Unlikely Activists Who Took on Silicon Valley—and Won," *New York Times*, Aug 14, 2018.

41. Russell Brandom, "Crucial Biometric Privacy Law Survives Illinois Court Fight," *The Verge*, Jan 26, 2019.

42. Colin Lecher, "New York City's Algorithm Task Force Is Fracturing," The Verge, Apr 15, 2019.

43. Joseph J. Lazzarotti et al., "State Law Developments in Consumer Privacy," *National Law Review*, Mar 15, 2019.

44. Morgan Sung, "'Mr. Zuckerberg' Explains the Internet to Elderly Senators," *Mashable*, Apr 10, 2018.

45. Brian Schatz (@Brianschatz) on Twitter, Dec 13, 2018, 3:37 P.M.

46. Art Raymond, "Utah Sen. Orrin Hatch Calls Foul on Media Coverage of Zuckerberg Hearing," *Deseret News*, Apr 12, 2018.

47. Avi Selk, "'There's So Many Different Things!': How Technology Baffled an Elderly Congress in 2018," *Washington Post*, Jan 2, 2019.

48. Subcommittee Hearing, "Extremist Content and Russian Disinformation Online: Working with Tech to Find Solutions," Committee on the Judiciary, Oct 31, 2017.

49. *Big Tech and Democracy: The Critical Role of Congress, Technology and Public Purpose Project, Belfer Center for Science and International Affairs*, Harvard Kennedy School, Apr 2019.

50. Sean Burch, "'Senator, We Run Ads': Hatch Mocked for Basic Facebook Question to Zuckerberg," *The Wrap*, Apr 10, 2018.

51. Eli Pariser, *The Filter Bubble: How the New Personalized Web Is Changing What We Read and How We Think* (London: Penguin Books, 2012).

52. Roger McNamee, "How to Fix Facebook: Make Users Pay for It," *Washing-*

ton Post, Feb 20, 2019; Maham Abedi, "A Facebook Subscription Fee? Tech Expert Says Yes, Mark Zuckerberg Isn't Too Sure," *Global News*, Apr 11, 2018.

53. Sam Dorman, "Josh Hawley: Tech Giants Have a Business Model Based on 'Taking Stuff from Us,'" *Fox News*, Aug 9, 2019.

54. Wikipedia, "List of Largest Internet Companies by Revenue and Market Capitalization," Wikipedia (webpage) https://en.wikipedia.org/wiki/list_of_largest_internet_companies.

55. Taylor Soper, "Report: Amazon Takes More Digital Advertising Market Share from Google—Facebook Duopoly," *Geekwire*, Feb 20, 2019.

56. "Gartner Says Worldwide IaaS Public Cloud Services Market Grew 31.3% in 2018," *Gartner Newsroom*, July 29, 2019.

57. Mike Isaac, "Mark Zuckerberg's Call to Regulate Facebook, Explained," *New York Times*, Mar 30, 2019. Carrie Mihalcik, "Sheryl Sandberg: Breaking Up Facebook Won't Fix Social Media," *Cnet*, May 17, 2019. Jon Porter, "Google's Sundar Pichai Snipes at Apple with Privacy Defense," *The Verge*, May 8, 2019. Romesh Ratnesar, "How Microsoft's Brad Smith Is Trying to Restore Your Trust in Big Tech," *Time*, Sep 9, 2019.

58. Isobel Hamilton, "Tim Cook Mounted His Most Stinging Attack Yet on Companies Like Facebook and Google that Hoard 'Industrial' Quantities of Data," *Business Insider*, Oct 24, 2018.

59. Zack Whittaker, "I Asked Apple for All My Data. Here's What Was Sent Back," *ZDNet*, May 24, 2018.

60. Sophia Yan, "China's New Cybersecurity Law Takes Effect Today, and Many Are Confused," CNBC, June 1, 2017.

61. Paul Mozur et al., "Apple Opening Data Center in China to Comply with Cybersecurity Law," *New York Times*, July 12, 2017.

62. "China Smartphone Market Share: By Quarter," Counterpoint Technology Market Research, Aug 28, 2019.

63. Mark Selden, "The Politics of Global Production: Apple, Foxconn and China's New Working Class," *Asia-Pacific Journal* (Japan Focus), Aug 8, 2013.

64. Jack Nicas and Paul Mozur, "In China Trade War, Apple Worries It Will Be Collateral Damage," *New York Times*, June 18, 2018.

65. Dipayan Ghosh, "Apple's Dangerous Market Grab in China," *New York Times*, July 18, 2017.

66. Zoe Schiffer, "'The Time Is Now to Have a Federal Privacy Bill,' Says Tim Cook," *The Verge*, Nov 22, 2019.

67. Lauren Feiner, "'Facebook Is the New Cigarettes,' Says Salesforce CEO," CNBC, Nov 14, 2018.

68. Ibid.

69. Steve Lohr, "How Top-Valued Microsoft Has Avoided the Big Tech Backlash," *New York Times*, Sep 8, 2019.

70. Olivia Solon, "As Tech Companies Get Richer, Is It 'Game Over' for Startups?" *The Guardian*, Oct 20, 2017.

71. Cass R. Sunstein, *#Republic: Divided Democracy in the Age of Social Media* (Princeton University Press, 2017).

72. Matthew Sheffield, "Facebook, Google and Twitter Admit Large-Scale Russian Infiltration," *Salon*, Nov 2, 2017.

73. David Ingram, "Facebook Says 126 Million Americans May Have Seen Russia-Linked Political Posts," Reuters, Oct 30, 2017.

74. Ian Bogost, "Another Day, Another Facebook Problem," *The Atlantic*, Sep 28, 2018.

75. Peter Walker, "Donald Trump Wins: Russian Parliament Bursts into Applause Upon Hearing Result," *The Independent*, Nov 9, 2016.

76. "Feinstein Criticizes Tech Companies for Not Addressing 'Cyber—Warfare,'" *Washington Post*, Nov 1, 2017.

77. Stephanie Murray, "Putin: I Wanted Trump to Win the Election," Politico, July 16, 2018.

Chapter 2

1. See Jeff Roberts, "The GDPR Is in Effect: Should U.S. Companies Be Afraid?," *Fortune*, May 24, 2018; Kristof van Quathem et al., "Google Fined €50 Million in France for GDPR Violation," *Inside Privacy* (blog), Covington & Burling, LLC, Jan 22, 2019; Axel Freiherr von Dem Bussche, "German Implementation of the GDPR," Taylor Wessing Partnerschaftsgesellschaft Mbb with Practical Law Data Privacy Advisor, Jul 2, 2018; Mar Masson Maack, "The Netherlands Premieres the First GDPR Fining Policy in the EU," Next Web, Mar 14, 2019; Laurent de Muyter, "Belgium Finalizes GDPR Implementation: A Practitioner's View," Privacy Tracker, International Association of Privacy Professionals, Sep 25, 2018; Paloma Bru and Paula Fernández-Longoria, "Spain Implements 'Urgent and Transitional Measures' on GDPR," *Pinsent Masons*, Aug 2, 2018.

2. UK Parliament (@Ukparliament) on Twitter, Feb 18, 2019.

3. Isabel Carvalho et al., "Are You Ready for Brazil's New Data Protection Law?" *Chronicle of Data Protection (blog), Hogan and Lovells*, Dec 27, 2018; Diego Fernandez, Argentina's New Bill on Personal Data Protection, Privacy Tracker, International Association of Privacy Professionals, Oct 2, 2018.

4. Michihiro Nishi, "Data Protection in Japan to Align with GDPR," Skadden, Arps, Slate, Meagher & Flom, Sep 24, 2018.; Doil Son and Sun Hee Kim, "Amendments to the Network Act Coming into Effect in 2019," *Lexology*, Jan 8, 2019.

5. Tim Gole et al., Australia's Privacy and Consumer Laws to Be Strengthened, *Lexology*, May 28, 2019; Justin Smulison, "Canada's Own 'GDPR' Now in Effect," *National Law Review*, Nov 7, 2018; Arindrajit Basu and Justin Sherman, "Key Global Takeaways from India's Revised Personal Data Protection Bill," *Lawfare* (blog), The Brookings Institution, Jan 23, 2020.

6. Daniel Rechtschaffen, "Why China's Data Regulations Are a Compliance Nightmare for Companies," *The Diplomat*, Jun 27, 2019; Emily Feng, "In China,

a New Call to Protect Data Privacy," NPR, Jan 5, 2020; Courtney Bowman, "A Primer on Russia's New Data Localization Law," *Proskauer*, Aug 27, 2015.

7. David Shepardson, "Trump Administration Working on Consumer Data Privacy Policy," Reuters, Jul 27, 2018; "Administration Discussion Draft: Consumer Privacy Bill of Rights Act of 2015," Obama White House Archives, National Archives and Records Administration.

8. "Congress Is Trying to Create a Federal Privacy Law," *The Economist*, Feb 28, 2019.

9. Drew Harwell, "Mass School Closures in the Wake of the Coronavirus Are Driving a New Wave of Student Surveillance," *Washington Post*, April 1, 2020.

10. Samuel Warren and Louis Brandeis, "The Right to Privacy," *Harvard Law Review*, vol. 4, no. 5, Dec 15, 1890.

11. "Privacy Online: Fair Information Practices in the Electronic Marketplace," Federal Trade Commission, May 2000.

12. Thomas Parke Hughes, *Networks of Power: Electrification in Western Society, 1880–1930* (Baltimore: Johns Hopkins University Press, 1993).

13. "Advanced Metering Infrastructure," conducted by the National Energy Technology Laboratory for the U.S. Department of Energy Office of Electricity Delivery and Energy Reliability, Feb 2008.

14. Mathew J. Morey, "Power Market Auction Design: Rules and Lessons in Market-Based Control for the New Electricity Industry," Edison Electric Institute, Sep 2001.

15. Mikhail Lisovich et al., "Inferring Personal Information from Demand-Response Systems," *IEEE Security & Privacy*, vol. 8, no. 1, Jan/Feb 2010.

16. Ibid.

17. Daniel Solove, "Will the United States Finally Enact a Federal Comprehensive Privacy Law?" TeachPrivacy, Apr 22, 2019.

18. Richard Thaler, "Some Empirical Evidence on Dynamic Inconsistency," *Economics Letters* 8, no. 3 (1981), 201–207.

19. Drazen Prelec and George Loewenstein, "Beyond Time Discounting," *Marketing Letters* 8, no. 1 (1997), 97–108.

20. Drazen Prelec and George Loewenstein, "Anomalies in Intertemporal Choice," chapter 33 in *Choices, Values, and Frames*, pp. 578–96 (*Cambridge University Press*, 2000); Drazen Prelec and George Loewenstein, "Anomalies in Intertemporal Choice: Evidence and an Interpretation," *Quarterly Journal of Economics*, 107, 2, pp. 573–97.

21. Alessandro Acquisti, "Privacy in Electronic Commerce and the Economics of Immediate Gratification," *Proceedings of the 5th ACM Conference on Electronic Commerce*, New York, 2004.

22. Leo Mirani, "Millions of Facebook Users Have No Idea They're Using the Internet, Quartz, Feb 9, 2015.

23. Lev Grossman, "Inside Facebook's Plan to Wire the World," *Time*, Dec 15, 2014.

24. Seth Feigerman, "Zuckerberg to Investors: If You Only Care about Money, Don't Buy Facebook," Mashable, Jan 28, 2015.

25. Mark Zuckerberg, "A Privacy-Focused Vision for Social Networking," Facebook, Mar 6, 2019.

26. Margaret Sullivan, "Mark Zuckerberg Claims that, at Facebook, 'the Future Is Private.' Don't Believe Him," *Washington Post*, May 5, 2019.

27. Rob Price, "Zuckerberg Says 'The Future Is Private,' But that Doesn't Mean What We Think," Business Insider, May 1, 2019.

28. Sundar Pichai, "Privacy Should Not Be a Luxury Good," *New York Times*, May 7, 2019.

29. Note, the assertion that Facebook deserves privacy regulation is likely true, as I have argued in an interview with Kara Swisher (see Eric Johnson, "Tech Is Now a Weapon for Propaganda and the Problem Is Way Bigger than Russia," *Vox*, Jan 31, 2018).

30. "Business Roundtable Redefines the Purpose of a Corporation to Promote 'an Economy That Serves All Americans,'" Business Roundtable, Aug 19, 2019.

31. See, for example, Jenny Anderson and Ephrat Livni, "Facebook's New Data Sharing Policies, Translated So a 13-Year-Old Can Understand," Quartz, May 2, 2018. The article notes that the company's terms of service are nonnegotiable, as is relatively consistent across the internet.

32. Natasha Lomas, "Facebook, Google Face First GDPR Complaints over 'Forced Consent,'" TechCrunch, May 25, 2018.

33. The German Federal Cartel Office has targeted precisely this "take-it-or-leave-it" situation in its recent suit against Facebook. See, for example, Natasha Lomas, "German Antitrust Office Limits Facebook's Data Gathering," TechCrunch, Feb 7, 2019.

34. "California Security Breach Notification Chart," Perkins Coie.

35. Leticia Miranda, "Thousands of Stores Will Soon Use Facial Recognition, and They Won't Need Your Consent," Buzzfeed, Aug 17, 2018.

36. Jenna Bitar and Jay Stanley, "Are Stores You Shop at Secretly Using Face Recognition on You?" American Civil Liberties Union, Mar 26, 2018.

37. "Starbucks Teams Up with Google to Bring Next-Generation WiFi Experience to Customers," Starbucks Stories & News, Jul 31, 2013.

38. Russell Brandom, "Facebook-Backed Lawmakers Are Pushing to Gut Privacy Law," The Verge, Apr 10, 2018.

39. Arnold Roosendahl, "Facebook Tracks and Traces Everyone: Like This!" *Tilburg Law School Legal Studies Research Paper Series* no. 03/2011, Dec 1, 2010.

40. Jane Wakefield, "Google's 'Secret Web Tracking Pages' Explained," BBC, Sep 5, 2019.

41. Ayden Férdeline, "Big Data, Not Big Brother: New Data Protection Laws and the Implications for Independent Media around the World," *CIMA Digital Report, National Endowment for Democracy*, June 6, 2019.

42. "What Are Cookies?" Knowledge Base, Indiana University (https://kb.iu.edu/d/agwm).

43. Brandy Zadrozny and Ben Collins, "Facebook Bans Ads from *The Epoch Times* After Huge Pro-Trump Buy," NBC News, Aug 22, 2019.

44. Bennett Cyphers, "Don't Play in Google's Privacy Sandbox," Electronic Frontier Foundation, Aug 30, 2019.

45. "Use of Cookies," *Scientific American* (www.scientificamerican.com/page/use-of-cookies/)

46. "The Facebook Pixel: A Piece of Code for Your Website that Lets You Measure, Optimize and Build Audiences for Your Ad Campaigns," Facebook for Business (www.facebook.com/business/learn/facebook-ads-pixel).

47. Natasha Lomas, "Europe's Top Court Sharpens Guidance for Sites Using Leaky Social Plug-Ins," TechCrunch, July 29, 2019.

48. See Allen St. John, "How Facebook Tracks You, Even When You're Not on Facebook," *Consumer Reports*, Apr 11, 2018; and Thomas Germain, "New 'off-Facebook Activity' Reveals How Company Tracks You All across the Web," *Consumer Reports*, Jan 29, 2020.

49. "Introducing Facebook's Audience Network," Facebook for Business, Apr 30, 2014.

50. "About Cookie Settings for Facebook Pixel," Facebook for Business.

51. "Do Not Track," Electronic Frontier Foundation (www.eff.org/issues/do-not-track).

52. "Turn 'Do Not Track' on or Off," Google Chrome Help (https://support.google.com/chrome/answer/2790761?co=GENIE.Platform%3DDesktop&hl=en).

53. John Corpuz, "Best Ad Blockers and Privacy Extensions," Tom's Guide, Aug 2, 2019.

54. "Privacy Badger," Electronic Frontier Foundation (www.eff.org/privacy badger).

55. Frederic Lardinois, "Google Wants to Phase Out Support for Third-Party Cookies in Chrome within Two Years," *TechCrunch*, Jan 14, 2020.

56. Dokyun Lee et al., "Advertising Content and Consumer Engagement on Social Media: Evidence from Facebook," *Management Science*, Jan 28, 2018.

57. Hooman Mohajeri Moghaddam et al., "Watching You Watch: The Tracking Ecosystem of Over-the-Top TV Streaming Devices," ACM Sigsac Conference on Computer and Communications Security, Nov 2019.

58. Arvind Narayanan (@Random_Walker) on Twitter, Sep 27, 2019.

59. Kevin Roose, "The Making of a YouTube Radical," *New York Times*, June 8, 2019.

60. Miller McPherson et al., "Birds of a Feather: Homophily in Social Networks," *Annual Review of Sociology*, 27 (2001), 415–44.

61. Nathan Eagle et al., "Inferring Friendship Network Structure by Using Mobile Phone Data," *Proceedings of the National Academy of Sciences*, Sep 8, 2009.

62. Linda Hollebeek et al., "Consumer Brand Engagement in Social Media:

Conceptualization, Scale Development and Validation," *Journal of Interactive Marketing*, 28 (2014), 149–65.

63. Xiao Han et al., "Alike People, Alike Interests? Inferring Interest Similarity in Online Social Networks," *Decision Support Systems*, 9 (2015), 92–106.

64. Natasha Lomas, "WhatsApp to Share User Data with Facebook for Ad Targeting—Here's How to Opt Out," TechCrunch, Aug 25, 2016.

65. Randy Alfred, "Dec. 8, 1993: Location, Location, Location," *Wired*, Dec 8, 2008.

66. Stewart Wolpin, "Commercial GPS Turns 25: How the Unwanted Military Tech Found Its True Calling," Mashable, May 25, 2014.

67. Alasdair Allan, "Got an IPhone Or 3G IPad? Apple Is Recording Your Moves," Radar, Apr 20, 2011.

68. Stephen Wicker, "The Loss of Location Privacy in the Cellular Age," Communications of the ACM, vol. 55, no. 8 (2012).

69. Rachel Levinson-Waldman, "Cellphones, Law Enforcement, and the Right to Privacy," Brennan Center for Justice at New York University School of Law, 2018.

70. Lily Hay Newman, "Carriers Swore They'd Stop Selling Location Data. Will They Ever?" *Wired*, Jan 9, 2019.

71. Kate Kaye, "Marketers Get on Board the Offline-to-Online Data Train," *Advertising Age*, May 20, 2014.

72. Suman Bhattacharya, "How Lowe's Is Using Pinterest to Build Customer Intent," Digiday, Oct 2, 2018.

73. Mark Bergen and Jennifer Surane, "Google and Mastercard Cut a Secret Ad Deal to Track Retail Sales," Bloomberg, Aug 30, 2018.

74. Emil Protalinski, "Android Passes 2.5 Billion Monthly Active Devices," VentureBeat, May 7, 2019

75. Natasha Lomas, "Google Tweaks Android Licensing Terms in Europe to Allow Google App Unbundling—for a Fee," TechCrunch, Oct 16, 2018.

76. Fred Campbell, "Google's Android Monopoly Isn't Free," *Forbes*, Aug 8, 2018.

77. Kelsey Sutton, "Google Is Collecting Your Data—Even When Your Phone Isn't in Use," *Adweek*, Aug 21, 2018.

78. Sundar Pichai, "Android Has Created More Choice, Not Less," Google Blog, July 18, 2018.

79. Chris Smith, "Let's Remember that Nobody Asked Google to Make Android Free," BGR, July 18, 2018.

80. Lee Matthews, "Facebook to Shut Down VPN App that Let It Spy on Users," *Forbes*, Feb 22, 2019.

81. John Bercovici, "Facebook Tried to Buy Snapchat for $3B in Cash. Here's Why," *Forbes*, Nov 13, 2013.

82. John Shinal, "Mark Zuckerberg Couldn't Buy Snapchat Years Ago, and Now He's Close to Destroying the Company," CNBC, July 14, 2017.

83. Emil Protalinski, "Facebook Is Stealing Employees from Everyone (Infographic)," ZDNet, Sep 28, 2011.

84. Tim Wu, "The Case for Breaking Up Facebook and Instagram," *Washington Post,* Sep 28, 2018.

85. "Facebook and the Cost of Monopoly," Stratechery, Apr 19, 2017.

86. Kevin Roose, "Facebook Emails Show Its Real Mission: Making Money and Crushing Competition," *New York Times,* Dec 5, 2018; Olivia Solon and Cyrus Farivar, "Mark Zuckerberg Leveraged Facebook User Data to Fight Rivals and Help Friends, Leaked Documents Show," NBC News, Apr 16, 2019.

87. See, for example, Charles Riley and Ivana Kottasová, "Europe Hits Google with a Third, $1.7 Billion Antitrust Fine," CNN Business, Mar 20, 2019; Aiofe White, "Facebook Is Latest Tech Company to Come Under EU's Antitrust Scrutiny," Bloomberg, July 2, 2019; "EU Antitrust Watchdog Considering Apple Probe: Vestager," Reuters, Mar 14, 2019.

88. See, for example, Brett Kendall, "FTC Antitrust Probe of Facebook Scrutinizes Its Acquisitions," *Wall Street Journal,* Aug 1, 2019; Nick Statt, "Google Is Facing an Imminent Antitrust Investigation from the US Justice Department," The Verge, May 31, 2019.

89. Ylan Mui, "Congress Is Getting Ready to Grill Top DOJ and FTC Officials about Being Too Lenient on Big Tech," CNBC, Sep 17, 2019.

90. "The Editorial Board: A $5 Billion Fine Won't Change Anything at Facebook. And There's a Bigger Problem," *Washington Post,* July 26, 2019.

91. Sara Fischer, "News Companies Take Big Tech Battle to Capitol Hill," Axios, June 11, 2019.

92. Natalie Nougayrède, "Zelensky's Election Proves Ukraine Is a Healthy Democracy. Putin Hates That," *The Guardian,* Apr 25, 2019.

93. John Schindler, "Why Vladimir Putin Hates Us," *Observer,* Nov 22, 2016.

94. Jeff Beer, "Facebook Says Sorry (Sort of) in Its Biggest Ever Ad Campaign," *Fast Company,* Apr 25, 2018.

95. Louis Menand, "Why Do We Care So Much about Privacy?" *The New Yorker,* June 11, 2018.

96. Sean Gallagher, "The Snowden Legacy, Part One: What's Changed, Really?" Ars Technica, Nov 21, 2018.

97. "Remarks by the President on Review of Signals Intelligence," Department of Justice, the Obama White House Archives, National Archives and Records Administration, Jan 17, 2014.

98. "Administration Discussion Draft: Consumer Privacy Bill of Rights Act of 2015," the Obama White House Archives, National Archives and Records Administration.

99. "Remarks by the President at the Cybersecurity and Consumer Protection Summit," White House Press office, *Stanford News,* Feb 13, 2015.

100. "In Wake of Equifax Data Breach, Blumenthal, Colleagues Introduce

Legislation to Hold Data Broker Industry Accountable," (press release). Office of Senator Richard Blumenthal, Sep 14, 2017.

101. Daniel Solove, "The Growing Problems with the Sectoral Approach to Privacy Law," TeachPrivacy, Nov 13, 2015.

102. Karl Bode, "AT&T Tricked Its Customers into Opposing Net Neutrality," Techdirt, July 19, 2017.

103. Cameron Kerry and Daniel Weitzner, "Rulemaking and Its Discontents: Moving from Principle to Practice in Federal Privacy Legislation," Brookings Institution, June 5, 2019.

104. Megan Brown et al., "President's Controversial Consumer Privacy Bill of Rights Act Informs Federal Privacy Dialogue, But Is Unlikely to Pass," Wiley Rein, Mar 2, 2015.

105. Natasha Singer, "Why a Push for Online Privacy Is Bogged Down in Washington," *New York Times*, Feb 29, 2016.

106. See "Equifax Data Breach Settlement," Federal Trade Commission, Sep 2019; Jon Brodkin, "Yahoo Tries to Settle 3-Billion-Account Data Breach with $118 Million Payout," Ars Technica, Apr 10, 2019; Ryan Francis, "MySpace Becomes Every Hackers' Space with Top Breach in 2016, Report Says," CSO, Feb 7, 2017; Michael Kassner, "Anatomy of the Target Data Breach: Missed Opportunities and Lessons Learned," ZDNet, Feb 2, 2015; Mike Isaac and Sheera Frenkel, "Facebook Security Breach Exposes Accounts of 50 Million Users," *New York Times*, Sep 28, 2018; "Marriott: Data on 500 Million Guests Stolen in 4-Year Breach," Krebs on Security, Nov 30, 2018; Tracy Kitten, "Wyndham Agrees to Settle FTC Breach Case," Bank Info Security, Dec 9, 2015; Brendan Koerner, "Inside the Cyberattack that Shocked the US Government," *Wired*, Oct 23, 2016; "Massive IRS Data Breach Much Bigger Than First Thought," CBS News, Feb 29, 2016.

107. W. Reece Hirsch and Ellie Chapman, "California Consumer Privacy Act Could Spell a Sea Change in US Privacy Law," Lexology, June 6, 2018.

108. "Privacy Is Gone, the Magic Bush," South Park Studios, Oct 29, 2014.

109. James Daniel, "Marketing Research System and Method for Obtaining Retail Data on a Real Time Basis," RPX Insight (https://insight.rpxcorp.com/pat/US4972504A).

110. Bo Pang and Lillian Lee, "Opinion Mining and Sentiment Analysis," *Foundations and Trends in Information Retrieval*, vol. 2, no. 1–2 (2008).

111. Chuck Bowen, "Review: Silicon Valley: Season One," *Slant*, Apr 1, 2014.

112. Virginia Smith et al., "Federated Multi-Task Learning," *Advances in Neural Information Processing Systems* 30 (2017).

113. Cynthia Dwork and Aaron Roth, "The Algorithmic Foundations of Differential Privacy," *Foundations and Trends in Theoretical Computer Science*, vol. 9, no. 3–4 (2014).

114. Craig Gentry, "A Fully Homomorphic Encryption Scheme," Stanford University, Sep 2009.

115. Dipayan Ghosh et al., "Economic Analysis of Privacy-Aware Advanced Metering Infrastructure Adoption," IEEE PES Innovative Smart Grid Technologies Conference, Jan 2012.

Chapter 3

1. "Charge of Discrimination, the Secretary, United States Department of Housing and Urban Development, on Behalf of Complainant, Assistant Secretary for Fair Housing and Equal Opportunity," *HUD v. Facebook*, Mar 28, 2019.

2. Clay Shirky, "The Political Power of Social Media: Technology, the Public Sphere, and Political Change," *Foreign Affairs*, vol. 90, no. 1 (Jan/Feb 2011).

3. See, for example, Thomas Poell and José Van Dijck, "Social Media and Activist Communication, the Routledge Companion to Alternative and Community Media," June 30, 2015; Merlyna Lim, "Clicks, Cabs, and Coffee Houses: Social Media and Oppositional Movements in Egypt, 2004–2011," *Journal of Communication*, vol. 62, no. 2 (Apr 2012).

4. Robin Mansell, "New Visions, Old Practices: Policy and Regulation in the Internet Era," *Journal of Media & Cultural Studies*, vol. 25, no. 1 (2011).

5. Nuala O'Connor, "Reforming the U.S. Approach to Data Protection and Privacy," Council on Foreign Relations, Jan 30, 2018.

6. Hemant Taneja, "The Era of 'Move Fast and Break Things' Is Over," *Harvard Business Review*, Jan 22, 2019.

7. Erin Griffith, "Will Facebook Kill all Future Facebooks? *Wired*, Oct 25, 2017.

8. Cecilia Kang and Sheera Frenkel, "Facebook Says Cambridge Analytica Harvested Data of Up to 87 Million Users," *New York Times*, April 4, 2018.

9. "Hearing on Protecting Consumer Privacy in the Era of Big Data, Subcommittee on Consumer Protection and Commerce of the Committee on Energy and Commerce," Feb 26, 2019.

10. Cat Zakrzewski, "The Technology 202: The Head of a Senate Tech Task Force Wants to Focus on Data Privacy," *Washington Post*, Aug 14, 2019.

11. "Potential Policy Proposals for Regulation of Social Media and Technology Firms," White Paper, U.S. Senator Mark R. Warner, July 23, 2018.

12. Kara Swisher, "Introducing the Internet Bill of Rights," *New York Times*, Oct 4, 2018.

13. Emily Stewart, "Josh Hawley's Bill to Limit Your Twitter Time to 30 Minutes a Day, Explained," Recode, July 31, 2019.

14. Will Oremus, "Google's Web of Confusion," *Slate*, Dec 11, 2018.

15. Dipayan Ghosh, "Facebook Is Allowing Politicians to Lie Openly. It's Time to Regulate," CNN Opinion, Oct 19, 2019.

16. Michael Grynbaum and Tiffany Hsu, "CNN Rejects 2 Trump Campaign Ads, Citing Inaccuracies," *New York Times*, Oct 3, 2019.

17. See, for example, Solon Barocas and Andrew Selbst, "Big Data's Disparate Impact," *California Law Review* vol. 104, no. 3 (2016).

18. Rich Caruana and Alexandru Niculescu-Mizil, "An Empirical Comparison of Supervised Learning Algorithms," *Proceedings of the 23rd International Conference on Machine Learning,* June 2006.

19. H. Altay Guvenir et al., "A Supervised Machine Learning Algorithm for Arrhythmia Analysis," Computers in Cardiology Conference, Sep 1997.

20. Alessandro Giusti et al., "A Machine Learning Approach to Visual Perception of Forest Trails for Mobile Robots, *IEEE Robotics and Automation Letters,* July 2016.

21. Yingying Wang et al., "Learning to Detect Frame Synchronization," International Conference on Neural Information Processing, 2013.

22. Yoshua Bengio and Yann Lecun, "Scaling Learning Algorithms towards AI," in *Large-Scale Kernel Machines* (Cambridge: MIT Press, 2007).

23. "About Targeting for Video Campaigns," YouTube Help; Caroline Donovan et al., "We Followed YouTube's Recommendation Algorithm Down the Rabbit Hole," Buzzfeed News, Jan 24, 2019.

24. James Zou and Londa Schiebinger, "AI Can Be Sexist and Racist—It's Time to Make It Fair," *Nature,* July 18, 2018.

25. Marianne Bertrand and Sendhil Mullainathan, "Are Emily and Greg More Employable than Lakisha and Jamal? A Field Experiment on Labor Market Discrimination," NBER Working Paper No. 9873, July 2003.

26. John Rawls, "Justice as Fairness," *Philosophical Review,* vol. 67, no. 2 (Apr 1958).

27. Jonathan Wolff, "Fairness, Respect, and the Egalitarian Ethos," *Philosophy & Public Affairs,* Apr 1998.

28. Kathleen Seiders and Leonard L. Berry, "Service Fairness: What It Is and Why It Matters," *Academy of Management Perspectives,* vol. 12, no. 2 (May 1998).

29. Daniel Kahneman et al., "Fairness and the Assumptions of Economics," *Journal of Business,* vol. 59, no. 4 (1986).

30. Jessica Guynn, "Google Photos Labeled Black People 'Gorillas,'" *USA Today,* July 1, 2015.

31. James Vincent, "Google 'Fixed' Its Racist Algorithm by Removing Gorillas from Its Image-Labeling Tech," The Verge, Jan 12, 2018.

32. Cory Doctorow, "Two Years Later, Google Solves 'Racist Algorithm' Problem by Purging 'Gorilla' Label from Image Classifier," Boing Boing, Jan 11, 2018.

33. Till Speicher et al., "Potential for Discrimination in Online Targeted Advertising," *Conference on Fairness, Accountability, and Transparency,* Feb 2018.

34. "Facebook Agrees to Sweeping Reforms to Curb Discriminatory Ad Targeting Practices," *American Civil Liberties Union,* Mar 19, 2019.

35. Rowland Manthorpe, "Google Directing People to Extreme Content and Conspiracy Theories, Sky News Finds," Sky News, June 30, 2019.

36. "Know Your Rights: Title VII of the Civil Rights Act of 1964," American Association of University Women.

37. "The Age Discrimination in Employment Act of 1967," U.S. Equal Employment Opportunity Commission.

38. “Fighting Discrimination in Employment Under the ADA,” Information and Technical Assistance on the Americans with Disabilities Act (www.ada.gov/employment.htm).

39. *Age Discrimination*, Sandra Norman-Eady, Chief Attorney, OLR Report, Dec 17, 2002.

40. Steven Willborn, “The Disparate Impact Model of Discrimination: Theory and Limits,” *American University Law Review vol.* 34, no. 3 (1985).

41. “Disparate-Impact Claims under the ADA,” American Bar Association, Apr 5, 2019.

42. Susan Grover, “The Business Necessity Defense in Disparate Impact Discrimination Cases,” *Georgia Law Review* vol. 30, no. 2 (1996).

43. Robert Garcia, “Garbage in, Gospel Out: Criminal Discovery, Computer Reliability, and the Constitution,” *UCLA Law Review* 1043, 1083 (1990–1991).

44. *Big Data: A Report on Algorithmic Systems, Opportunity, and Civil Rights* (Washington, D.C.: Executive office of the President, May 2016).

45. Pauline T. Kim, “Data-Driven Discrimination at Work,” *William & Mary Law Review* vol. 857 (2016–2017).

46. Faisal Kamiran and Toon Calders, “Classification with No Discrimination by Preferential Sampling,” Benelearn, May 27–28, 2010.

47. Elizabeth E. Joh, “Feeding the Machine: Policing, Crime Data, & Algorithms,” *William and Mary Bill of Rights Journal*, 2017–2018.

48. “When Digital Dust Is Gathered, Constellation May Be Muddled,” *National Public Radio*, Apr 13, 2013.

49. John Aldrich, “Correlations Genuine and Spurious in Pearson and Yule,” *Statistical Science*, vol. 10, no. 4 (1995).

50. Eric Seigel, “The Real Problem with Charles Murray and ‘The Bell Curve,’” *Scientific American*, Apr 12, 2017.

51. Ralf Klinkenberg and Thorsten Joachims, “Detecting Concept Drift with Support Vector Machines,” *ICML ‘00 Proceedings of the Seventeenth International Conference on Machine Learning*, June 29, 2000.

52. Raymond S. Nickerson, “Confirmation Bias: A Ubiquitous Phenomenon in Many Guises,” *Review of General Psychology*, Jun 1, 1998.

53. Karen Hao, “This Is How AI Bias Really Happens—and Why It’s So Hard to Fix,” *MIT Technology Review*, Feb 4, 2019.

54. Jeffrey Dastin, “Amazon Scraps Secret AI Recruiting Tool that Showed Bias Against Women,” Reuters, Oct 9, 2018.

55. Jessica Guynn, “Facebook Vows to Stop Ad Discrimination against African-Americans, Women and Older Workers,” *USA Today*, Mar 20, 2019.

56. Olivia Solon, “YouTube’s ‘Alternative Influence Network’ Breeds Rightwing Radicalisation, Report Finds,” *The Guardian*, Sep 18, 2018.

57. Julia Angwin et al., “Machine Bias,” ProPublica, May 23, 2016.

58. Anupam Datta et al., “Proxy Non-Discrimination in Data-Driven Systems,” arXiv, July 25, 2017.

59. Dexter Thomas, "Facebook Tracks Your 'Ethnic Affinity'—Unless You're White," Vice, Nov 16, 2016.

60. "Facebook Lets Advertisers Control shich Ethnicities See Their Ads," *Vice*, Oct 28, 2016.

61. Steven Melendez, "Facebook Promises to Nix Its Tools for Discriminatory Ad Targeting," *Fast Company*, July 24, 2018.

62. Katie Benner et al., "Facebook Engages in Housing Discrimination with Its Ad Practices, U.S. Says," *New York Times*, Mar 28, 2019.

63. Victor Luckerson, "Here's Why Facebook Won't Put Your News Feed in Chronological Order," *Time*, July 9, 2015.

64. John Maxwell Hamilton, "In a Battle for Readers, Two Media Barons Sparked a War in the 1890s," *National Geographic*, Apr 16, 2019.

65. Kim Peterson, "Whiskey That Tastes Like 'A Burning Hospital,'" CBS News, Oct 20, 2014.

66. Thomas Eisenmann, "Internet Companies' Growth Strategies: Determinants of Investment Intensity and Long-Term Performance," *Strategic Management Journal*, Oct 30, 2006.

67. Gilles Deleuze and Felix Guattari, *A Thousand Plateaus: Capitalism and Schizophrenia*, translation by Brian Massumi (University of Minnesota Press, 1987).

68. Bruno Latour, "Networks, Societies, Spheres: Reflections of an Actor-Network Theorist," *International Journal of Communication*, 2011; Michel Callon, Actor-Network Theory—The Market Test, *Sociological Review*, May 1, 1999; John Law and John Hassard (eds.), *Actor Network Theory and After* (Oxford: Blackwell, 1999).

69. See, for example, "Advertising and Disclaimers," Federal Election Commission.

70. "SPJ Code of Ethics," Society of Professional Journalists.

71. Gordon Pennycook et. al., "Prior Exposure Increases Perceived Accuracy of Fake News," *Journal of Experimental Psychology*, Dec 2018.

72. Soroush Vosoughi et al., "The Spread of True and False News Online," *Science*, vol. 359, no. 6380 (Mar 9, 2018).

73. Cass Sunstein, *#Republic: Divided Democracy in the Age of Social Media*, (Princeton University Press, 2017).

74. Helen Nissenbaum, "A Contextual Approach to Privacy Online," *Daedalus vol.* 140, no. 4 (2011), 32–48.

75. Liz Entman, "Study of Google Data Collection Comes Amid Increased Scrutiny over Digital Privacy," Phys.org, Nov 2 2018.

76. B. Mahadevan, "Business Models for Internet-Based E-Commerce: An Anatomy," *California Management Review*, July 1, 2000.

77. Robbin Lee Zeff and Bradley Aronson, *Book Advertising on the Internet* (Hoboken, NJ: John Wiley & Sons, 1999).

78. Joseph Cox, "I Gave a Bounty Hunter $300. Then He Located Our Phone," *Motherboard (blog)*, Vice, January 8, 2019.

79. "Charge of Discrimination, the Secretary, United States Department of Housing and Urban Development, on Behalf of Complainant, Assistant Secretary for Fair Housing and Equal Opportunity," *HUD v. Facebook*, Mar 28, 2019.

80. Adam Mosseri, "Bringing People Closer Together," *Facebook Newsroom*, Jan 11, 2018.

81. Michael Grothaus, "Facebook's 'Meaningful Interaction' News Feed Change Is a Failure," *Fast Company*, Mar 18, 2019.

82. Kelly Wynne, "How to See Your Secret Instagram 'Ad Interests,'" Jun 7, 2019.

83. Kari Paul, "Facebook Tracks Everything from Your Politics to Ethnicity—Here's How to Stop It," MarketWatch, Jan 19, 2019.

84. Natasha Lomas, "YouTube Under Fire for Recommending Videos of Kids with Inappropriate Comments," TechCrunch, Feb 18, 2019.

85. Alexis Madrigal, "Before It Conquered the World, Facebook Conquered Harvard," *The Atlantic*, Feb 4, 2019.

86. Jessica Salter, "Airbnb: The Story Behind the $1.3bn Room-Letting Website," *The Telegraph*, Sep 7, 2012.

87. Chris Welch, "Gmail Is 10 Years Old Today," *The Verge*, Apr 1, 2014.

88. Olivia Solon, "Teens Are Abandoning Facebook in Dramatic Numbers, Study Finds," *The Guardian*, June 1, 2018.

89. Sandra Wachter, "Affinity Profiling and Discrimination by Association in Online Behavioural Advertising," *Berkeley Technology Law Journal*, vol. 35, no. 2 (2020).

Chapter 4

1. Jay Shambaugh et al., "The State of Competition and Dynamism: Facts about Concentration, Start-Ups, and Related Policies," Brookings Institution, June 13, 2018.

2. Jacob M. Schlesinger et al., "Tech Giants Google, Facebook and Amazon Intensify Antitrust Debate," *Wall Street Journal*, June 12, 2019.

3. Louise Matsakis, "Break Up Big Tech? Some Say Not So Fast," *Wired*, June 7, 2019.

4. Matt Stevens, "Elizabeth Warren on Breaking Up Big Tech," *New York Times*, June 26, 2019.

5. Philip Verveer, "Platform Accountability and Contemporary Competition Law: Practical Considerations," Shorenstein Center on Media, Politics and Public Policy, Harvard Kennedy School, Nov 20, 2018.

6. Liza Gormsen and Jose Llanos, "Facebook's Anticompetitive Lean in Strategies," *SSRN*, June 17, 2019.

7. Sara Salinas, "Amazon's Ad Business Will Steal Market Share from Google This Year, Says eMarketer," CNBC, Feb 20, 2019.

8. Seth Fiegerman, "Amazon Has Its First $200 Billion Sales Year, But Growth Is Slowing," CNN Business, Jan 31, 2019.

9. Steven Melendez and Alex Pasternack, "Here Are the Data Brokers Quietly Buying and Selling Your Personal Information," *Fast Company*, Mar 2, 2019.

10. Jordan Novet, "Amazon Web Services Reports 45 Percent Jump in Revenue in the Fourth Quarter," CNBC, Jan 31, 2019.

11. Amy Gesenhues, "Snap Topped $1 Billion in Revenue in 2018, Stabilized User Base in Fourth Quarter," MarTech Today, Feb 6, 2019.

12. Kerry Flynn, "Media Buyers Praise Twitter's Ad Platform as Simple and Reliable," Digiday, Dec 6, 2018.

13. Alex Johnson, "Verizon Says Oath Is Dead. Meet Verizon Media Group," NBC News, Dec 18, 2018.

14. Tim Peterson, "5 Things We Learned about AT&T's Media and Advertising Business in 2018," Digiday, Dec 31, 2018.

15. Jordan Kramer, "Meet the Top 10 Digital Ad Spenders in 2017 So Far," Pathmatics, Sep 14, 2017.

16. Claire Cain Miller, "How Jean Tirole's Work Helps Explain the Internet Economy," *New York Times*, Oct 15, 2014.

17. Lawrence Lessig, "Do You Floss?" *London Review of Books*, Aug 2004.

18. Dan Gallagher, "Data Really Is the New Oil," *Wall Street Journal*, Mar 9, 2019.

19. Dipayan Ghosh and Ben Scott, "Digital Deceit II: A Policy Agenda to Fight Disinformation on the Internet," New America and the Shorenstein Center on Media, Politics and Public Policy, Harvard Kennedy School, Sep 24, 2018.

20. Paul Hitlin and Lee Rainie, "Facebook Algorithms and Personal Data," Pew Research Center, Jan 16, 2019.

21. See, for example, Patrick Rey and Joseph Stiglitz, "Vertical Restraints and Producers' Competition," NBER, May 1988.

22. Michelle Castillo, "Mark Zuckerberg Hints that Facebook Has Considered a Paid Version," CNBC, Apr 10, 2018.

23. Sara Fischer, "Sandberg: There'd Be a 'Paid Product' If Users Wanted No Ads," Axios, Apr 5, 2018.

24. Ibid.

25. "Bundeskartellamt Prohibits Facebook from Combining User Data from Different Sources," Bundeskartellamt, Feb 7, 2019.

26. Sara Germano, "Facebook Wins Appeal against German Data-Collection Ban," *Wall Street Journal*, Aug 26, 2019.

27. Allan Weis, "Commercialization of the Internet," *Internet Research*, vol. 20, no. 4 (Aug 17, 2010).

28. N. Economides, "The Telecommunications Act of 1996 and Its Impact," *NYU Stern School of Business*, 1998.

29. Brian Mccullough, "An Eye-Opening Look at the Dot-Com Bubble of 2000—And How It Shapes Our Lives Today," TED, Dec 4, 2018.

30. Manish Agarwal and David Round, "The Emergence of Global Search Engines: Trends in History and Competition," *Competition Policy International*, vol. 7, no. 1 (2011).

31. Sam Thielman, "Yahoo Is Not Alone: Six Failed Tech Companies and How They Fell," *The Guardian*, July 25, 2016.

32. Christopher Yoo, *The Dynamic Internet: How Technology, Users, and Businesses Are Transforming the Network* (Washington, DC: AEI Press, 2012).

33. Ruqayyah Moynihan et al., "25 Giant Companies that Are Bigger than Entire Countries," *Business Insider*, July 25, 2018.

34. J. Clement, "Primary E-Mail Providers According to Consumers in the United States as of January 2017, by Age Group," Statista, Oct 23, 2018.

35. J. Clement, "Share of Search Queries Handled by Leading U.S. Search Engine Providers as of July 2019," Statista, Oct 16, 2019.

36. Arne Holst, "Mobile Operating Systems' Market Share Worldwide from January 2012 to July 2019," Statista, Sep 13, 2019.

37. Jun-Sheng Li, "How Amazon Took 50% of the E-Commerce Market and What It Means for the Rest of Us," TechCrunch, Feb 27, 2019.

38. J. Clement, "Most Popular Social Network Websites in the United States in August 2019, Based on Share of Visits," Statista, Sep 9, 2019.

39. J. Clement, "Most Popular Mobile Messaging Apps in the United States as of June 2019, by Monthly Active Users (in Millions)," Statista, Nov 12, 2019.

40. David P. Reed, "That Sneaky Exponential—Beyond Metcalfe's Law to the Power of Community Building," *Context Magazine*, 1999.

41. Dean Eckles et al., "Estimating Peer Effects in Networks with Peer Encouragement Designs," *Proceedings of the National Academy of Sciences*, July 5, 2016.

42. See, for example, Jon Markman, "The Amazon Era: No Profits, No Problem," *Forbes*, May 23, 2017; Erik Sherman, "4 Reasons YouTube Still Doesn't Make A Profit," CBS News, May 27, 2015.

43. Yochai Benkler et al., *Network Propaganda* (Oxford University Press, 2018).

44. Hunt Allcott, Luca Braghieri, Sarah Eichmeyer, Matthew Gentzkow, "The Welfare Effects of Social Media," NBER Working Paper No. 25514, Nov 2019.

45. Nobuo Okishio, "Technical Change and the Rate of Profit," *Kobe University Economic Review*, 1961.

46. Boris Marjanovic, "An (Old) Way to Pick (New) Stocks," Seeking Alpha, Jan 26, 2016.

47. "Shutting Down Partner Categories," Facebook Newsroom, Mar 28, 2018.

48. Marty Swant, "Snapchat and Nielsen Partner to Integrate Offline Data for Audience Targeting," *Adweek*, Jul 19, 2018.

49. "Monopoly by the Numbers, Monopoly Basics," Open Markets Institute (https://openmarketsinstitute.org/monopoly-basics/).

50. Langdon Winner, "Technologies as Forms of Life," in *Ethics and Emerging Technologies*, edited by Ronald L. Sandler (London: Palgrave Macmillan, 2014).

51. Jeremy Greenwood et al., "Long-Run Implications of Investment-Specific Technological Change," *American Economic Review*, June 1997.

52. On Exclusive Dealing, see Nicholas Economides, :What Google Can Learn from Microsoft's Antitrust Problems," *Fortune*, July 17, 2016.

53. "Anticompetitive Practices," Organisation for Economic Co-operation and Development, Jan 3, 2002.

54. John Howells, "The Response of Old Technology Incumbents to Technological Competition—Does the Sailing Ship Effect Exist?" *Journal of Management Studies*, Feb 17, 2003.

55. See, for example, Ryan Calo, "Digital Market Manipulation," *George Washington Law Review* vol. 82, no. 4 (2014).

56. Lesley Fair, "FTC's $5 Billion Facebook Settlement: Record-Breaking and History-Making," *Federal Trade Commission*, July 24, 2019.

57. "Why the Consumer Welfare Standard Remains the Best Guide for Promoting Competition," *Competition Policy International*, Jan 27, 2019.

58. "Benefits of Competition and Indications of Market Power," Issue Brief, *Council of Economic Advisers, the Obama White House Archives, National Archives and Records Administration*, Apr 2016.

59. R. Bork, *The Antitrust Paradox* (New York: *Free Press*, 1978).

60. J. Stiglitz, "Towards a Broader View of Competition Policy," Roosevelt Institute, June 2017.

61. "Competition and Monopoly: Single-Firm Conduct Under Section 2 of the Sherman Act: Chapter 2," Antitrust Division, U.S. Department of Justice.

62. See, for example, Matt Binder, "Google Hit with $1.7 Billion Fine for Anticompetitive Ad Practices," Mashable, Mar 20, 2019; Jason Del Rey, "Amazon May Soon Face an Antitrust Probe. Here Are 3 Questions the FTC Is Asking about It," Recode, June 4, 2019.

Chapter 5

1. Jean-Jacques Rousseau, *Discourse on the Origin and Basis of Inequality Among Men* (The Second Part) (Holland: Marc-Michel Rey, 1755).

2. "GDPR One Year On," *BBC Click*, June 13, 2019.

3. This is indeed where a famous case in Spain concerning Google Search records landed. See Robert C. Post, "Data Privacy and Dignitary Privacy: Google Spain, the Right to Be Forgotten, and the Construction of the Public Sphere," *Duke Law Journal* vol. 67, no. 5 (Feb 2018): 981–1072; Eleni Frantziou, "Further Developments in the Right to Be Forgotten: The European Court of Justice's Judgment in Case C-131/12, Google Spain, Sl, Google Inc V Agencia Espanola De Proteccion De Datos," *Human Rights Law Review* vol. 14, no. 4 (2014), 761–78.

4. Tim Wu, *The Curse of Bigness: Antitrust in the New Gilded Age* (New York: Columbia Special Reports, 2018); B. Lynn, "A Missing Key to Prosperity, Oppor-

tunity, and Democracy," Demos, 2013; Elizabeth Warren, "Here's How We Can Break Up Big Tech," *Team Warren Medium*, Mar 8, 2019; J. Abbruzzese, "Lindsey Graham Pushes Zuckerberg on Competition (Or Lack Thereof)," *NBC News*, April 10, 2018.

5. C. Shapiro, "Antitrust in a Time of Populism," *International Journal of Industrial Organization* (Feb 27, 2018); F. Easterbrook, "The Limits of Antitrust," *Texas Law Review* (Aug 1984).

6. J. Furman, "Note on Market Concentration," *OECD*, June 7, 2018.

7. Richard Blumenthal and Tim Wu, "What the Microsoft Antitrust Case Taught Us," *New York Times*, May 18, 2019.

8. Chris Hughes, "It's Time to Break Up Facebook," *New York Times*, May 9, 2019.

9. Astead W. Herndon, "Elizabeth Warren Proposes Breaking Up Tech Giants Like Amazon and Facebook," *New York Times*, Mar 8, 2019.

10. "Sen. Cruz Delivers Introductory Remarks as Chairman of Big Tech Censorship Judiciary Hearing," Senator Ted Cruz on YouTube.

11. Robert Reich, "Elizabeth Warren Is Right—We Must Break Up Facebook, Google and Amazon," *The Guardian*, Mar 10, 2019.

12. David Mclaughlin and Ben Brody, "Breaking Up Facebook Is Easier Said Than Done," Bloomberg, May 10, 2019.

13. Nick Clegg, "Breaking Up Facebook Is Not the Answer," *New York Times*, May 11, 2019.

14. Elizabeth Warren, "Here's How We Can Break Up Big Tech," *Team Warren Medium*, Mar 8, 2019.

15. "Monopolization Defined," Federal Trade Commission (www.ftc.gov/tips-advice/competition-guidance/guide-antitrust-laws/single-firm-conduct/monopolization-defined).

16. Thomas Graf and Alexander Waksman, "Antimonopoly & Unilateral Conduct," European Union, *Global Competition Review*.

17. Lina Khan, "Amazon's Antitrust Paradox," *Yale Law Journal*, vol. 126 (2016–2017).

18. Feng Zhu and Marco Iansiti, "Why Some Platforms Thrive and Others Don't," *Harvard Business Review* (Jan–Feb 2019).

19. See, for example, Mark Sullivan, "Facebook Should Never Have Been Allowed to Buy Instagram, Silicon Valley Rep Says," *Fast Company*, Jan 25, 2019.

20. See, for example, Ilker Koksal, "Snapchat's Plans for 2020: Bitmoji TV Goes Live," *Forbes*, Feb 4, 2020.

21. Makena Kelly, "Facebook Plans to Tie Itself Together as Regulators Debate Tearing It Apart," The Verge, Mar 7, 2019.

22. Ezra Klein, "Facebook Is a Capitalism Problem, Not a Mark Zuckerberg Problem," Recode, May 10, 2019.

23. Dipayan Ghosh, "A New Digital Social Contract Is Coming for Silicon Valley, *Harvard Business Review*, Mar 27, 2019.

24. Federal Trade Commission, "FTC Issues Statement of Principles Regarding Enforcement of FTC Act as a Competition Statute," Aug 13, 2015. Robert Bork, *the Antitrust Paradox* (New York: Free Press, 1978).

25. S. Pozen, "Acting Assistant Attorney General Sharis A. Pozen Speaks at the American Bar Association 2011 Antitrust Fall Forum," Nov 17, 2011.

26. Georgia Wells and Deepa Seetharaman, "Snap Detailed Facebook's Aggressive Tactics in 'Project Voldemort' Dossier," *Wall Street Journal,* Sep 24, 2019.

27. John Kenneth Galbraith, *The Affluent Society* (New York: Houghton Mifflin, 1958).

28. Dipayan Ghosh, "Banning Micro-Targeted Political Ads Won't End the Practice," *Wired,* Nov 22, 2019.

29. "Political Campaigning, Advertising Guidelines," Pinterest (https://policy.pinterest.com/en/advertising-guidelines).

30. Patience Haggin, "Google to Restrict Political Ad Targeting on Its Platforms," *Wall Street Journal,* Nov 20, 2019; Emily Glazer, "Facebook Weighs Steps to Curb Narrowly Targeted Political Ads," *Wall Street Journal,* Nov 21, 2019.

31. Daniel Kreiss and Shannon Mcgregor, "Technology Firms Shape Political Communication: The Work of Microsoft, Facebook, Twitter, and Google with Campaigns During the 2016 U.S. Presidential Cycle," *Political Communication,* vol. 35, no. 2 (Oct 26, 2017).

32. Ramsi Woodcock, "Advertising as Monopolization in the Information Age," *Antitrust Chronicle,* Competition Policy International, Apr 2019.

33. See, for example, Nicholas Diakopoulos, "Accountability in Algorithmic Decision Making," *Communications of the ACM,* vol. 59, no. 2 (2016), 56–62.

34. David Gunning, *Explainable Artificial Intelligence (XAI),* Defense Advanced Research Projects Agency, Nov 2017.

35. Davey Alba, "Ahead of 2020, Facebook Falls Short on Plan to Share Data on Disinformation," *New York Times,* Sep 29, 2019.

36. Tom Wheeler, "How to Monitor Fake News," *New York Times,* Feb 20, 2018.

37. Nicholas Diakopoulos et al., "Principles for Accountable Algorithms and a Social Impact Statement for Algorithms," Fairness, Accountability, Transparency in Machine Learning (www.fatml.org/resources/principles-for-accountable-algorithms).

38. National Institute of Standards and Technology, "Cybersecurity Framework," U.S. Department of Commerce.

39. Yochai Benkler, "Don't Let Industry Write the Rules for AI," *Nature* vol. 569, no. 161 (2019).

40. Agreement Containing Consent Order, File 092 3184, *in the Matter of Facebook, Inc.*, Federal Trade Commission (Facebook), Nov 29, 2011; Agreement Containing Consent Order, File 122 3237, *in the Matter of Google, Inc.*, Federal Trade Commission (Google), Sep 4, 2014.

41. "PWC Had Cleared Facebook's Privacy Practices in Leak Period," Reuters, Apr 19, 2018.

42. Jeremy Merrill and Ariana Tobin, "Facebook Moves to Block Ad Transparency Tools—Including Ours," ProPublica, Jan 28, 2019.

43. Dipayan Ghosh and Ben Scott, "Digital Deceit II: A Policy Agenda to Fight Disinformation on the Internet," New America, Sep 24, 2018.

44. Andrew Rice, "This Is Ajit Pai, Nemesis of Net Neutrality," *Wired*, May 16, 2018.

45. Michael Nunez, "Former Facebook Workers: We Routinely Suppressed Conservative News," Gizmodo, May 9, 2016.

46. James Vincent, "Mark Zuckerberg Is Meeting with Glenn Beck to Talk about Facebook Bias," The Verge, May 16, 2016.

47. Nick Bostrom, "Existential Risks: Analyzing Human Extinction Scenarios and Related Hazards," *Journal of Evolution and Technology*, 2002.

48. Mancur Olson, *The Logic of Collective Action* (Harvard University Press, 1971).

49. Christian Joppke, "The Crisis of the Welfare State, Collective Consumption, and the Rise of New Social Actors," *Berkeley Journal of Sociology*, vol. 32 (1987).

50. "Our Members," *Internet Association* (https://internetassociation.org/our-members/).

51. Dipayan Ghosh, "Facebook's Oversight Board Is Not Enough," *Harvard Business Review*, Oct 16, 2019.

52. Alistair Barr, "Facebook's China Argument Revealed in Zuckerberg's Hearing Notes," Bloomberg, Apr 10, 2018.

53. David Collingridge, *The Social Control of Technology* (London: Palgrave Macmillan, 1981).

The Case for Radical Reform

1. Casey Newton, "Tech Giants Are Finding Creative Ways to Use Our Data to Fight the Coronavirus," The Verge, April 7, 2020.

2. Danielle Abril, "Former Google CEO: The Coronavirus Pandemic Will Make Big Tech Even Bigger," *Fortune*, April 7, 2020.

3. Adrien Abecassis, Dipayan Ghosh, and Jack Loveridge, "La crise du coronavirus ébranle aussi l'idée de démocratie et de liberté," *Le Monde*, March 26, 2020.

Index

Aadhaar data breach, 122
academia: intellectual obscurantism in, 79; knowledge-keeping paradigm of, 23–26
access-to-data rules, 68, 206
Acquisti, Alessandro, 78
actor-network theory, 150
Acxiom (data broker), 207
adblock software, 98–99
addiction, to social media, 33, 54–55, 59, 185
advertising: adblock software, 98–99; Facebook-Google-Amazon market dominance in, 164–170, 186, 187–189; technological advances, effects on, 147–148; terms of service and, 85, 86. *See also* attention-data currency
advertising, political: Facebook and, 95, 130, 226, 233, 234; Honest Ads Act, 36, 234, 241; platforms allow or block, 226–227; regulatory protections regarding, 36, 130, 226–229, 233–234, 241, 250; transparency in, 226–229, 233, 234, 241. *See also* election interference
advertising, regulatory protections regarding, 13–14, 138–140, 161; digital social contract and, 200, 202, 204, 209–210, 219, 221, 224; political ads, 36, 130, 226–229, 233–234, 241, 250
advertising, targeted, 13–14, 17, 59–60, 62–63, 186–187; Audience Network, 81, 96, 102, 194; civil rights law and, 138–139; click-through rates, 99–100, 108, 138, 141; cookies and pixels, 94, 96–98; criticism of, by Apple, 51;

advertising, (*cont.*)
discrimination in, 136, 138, 141, 145, 155–156; engagement data, 100, 194; Facebook feed, 41, 60, 79, 148, 187; location data, 105; marketing analysis, 108; "meaningful social interaction" claims, 157–158; Partner Categories, 189–190; social-graph data, 102–104; transaction data and, 107–108; web browsing and mobile-use data, 94–99. *See also* algorithms, recommendation and content curation; Amazon, advertising and; attention-data currency; Cambridge Analytica incident; data inferences; Facebook, advertising and; Google, advertising and

advertising revenue: Amazon, 48, 49, 164–165, 166–167, 180, 186; Apple, 51; Facebook, 48, 49, 56, 57, 178, 180, 186; Google, 48, 49, 165, 178, 180, 186, 244; political ads, 226–227; smaller companies, 189; terms of service and, 86; top companies, compared, 49, 178, 180, 186, 187; YouTube, 49, 59. *See also* revenue and market share rankings

age demographics, platform use and, 147, 160

Airbnb, 45, 47, 48

Alcott, Hunt, 185

algorithmic discrimination, 43, 134–161; confirmation bias, 142–143; disparate impact cases, 136, 137–139, 142, 156, 158, 160; drift problem, 142–143, 144; facial-recognition technology, 38, 92–93; fairness and justice arguments, 134–136; high-value audiences, 159–160; "meaningful social interaction" claims, 157–159; print media, compared, 146–147; proxies in, 144–145; racism and, 134, 135, 144, 145; radical commercialization of decisions, 148–154; regulatory protections against, 161, 208, 223–227, 229–230, 232–233; as rhizomatic, 149–150, 151, 153, 210; training data, 131, 139–142, 143; ubiquity of, 154–160. *See also* behavioral profiling; data inferences

algorithms, 58, 127–161, 187, 220, 224; creation and design of, 130–134, 139–142, 143, 160–161, 230–233; engagement data and, 99–100, 194–196; ethical AIs, 230–231; privacy engineering and, 124–125; regulatory oversight of, 231–233; transparency and, 232–233

algorithms, recommendation and content curation, 17, 127–128, 132–134; Facebook feed, 58, 60, 144, 148, 187, 233; YouTube "rabbit hole" effect, 10, 33, 59, 134, 136, 202. *See also* advertising, targeted; algorithms; content production and curation

Allan, Alasdair, 104

Allsop, Jon, 12

Alphabet (Google parent company), 56. *See also* Google

Amazon, 29, 48; algorithmic hiring discrimination, 143; Bezos and, 12, 28–29; facial-recognition technology and, 93; network effect and, 183; service offerings, 165; vertical integration, 193; Whole Foods, 222. *See also* Facebook-Google-Amazon market dominance

Amazon, advertising and, 164–170, 178–179; as challenge to Facebook and Google market share, 164–165, 166–170, 178, 186; late entrance

by, 164–166; revenue from, 48, 49, 164–165, 180, 186. *See also* advertising
Amazon, as e-commerce monopoly. *See* Facebook-Google-Amazon market dominance
Amazon, revenue and market share, 45, 56; from advertising, 48, 49, 164–165, 166–167, 180, 186; growth, 164–165, 167; online video viewing, 175; retail online sales, 177. *See also* revenue and market share rankings
Amazon Prime Video, 176
Amazon Web Services, 48
American Civil Liberties Union (ACLU), 92–93, 136
Android mobile OS, 108–111
anticompetitive practices, 17–20, 43, 55, 85, 108, 164; digital social contract as response to, 201, 221; innovation, copying, 51–52, 112, 129, 194, 197, 218–219; regulatory inquiries into, 112–113, 196–197, 201, 221, 250; smaller rivals, 189–190, 191. *See also* digital social contract, competition issues; Facebook-Google-Amazon market dominance; monopolies
anticonservative bias, 33, 34–35, 44
antidiscrimination law, 137, 143, 144, 145. *See also* algorithmic discrimination; discrimination
antitrust law and regulatory framework, 20, 33–40, 55, 163–164, 196–198, 250; Antitrust Division, Justice Department, 215, 221–222; break-up policies, 164, 197, 214–215, 217, 219, 221, 243; Cartel Office (Bundeskartellamt) suit, 179–181; competition regulation inquiries, 1–2, 35–40, 112–113, 116, 179–181, 216–217; consumer-harm standards, 212, 221; enforcement, 212–213, 221–222; natural monopolies, 190–191, 213, 215–220. *See also* digital social contract, competition issues; monopolies; regulatory protections
Apple: China and, 52, 53–54, 83, 85, 243–244; Cook and, 50–52, 54, 83–85; data and, 50–51, 54, 83; Facebook criticism by, 51, 83–84; industrial obscurantism by, 83–85; iOS, 108, 110; location data and, 104; revenue and market share, 45–46, 50–55, 56, 128, 177
Apple Maps, 174
Arab Spring, 21, 41
Aral, Sinan, 152
Archive of Our Own, 171
Ardern, Jacinda, 14
Argentina, 38, 66
artificial intelligence (AI) systems, 92, 226, 230–231. *See also* algorithms
AT&T, 166
attention-data currency, 167–181, 193–196, 216; Cartel Office (Bundeskartellamt) suit, 179–181; engagement maximization and, 58–59, 195; Facebook and, 96–98, 157, 170–171, 178, 179; Facebook-Google-Amazon market dominance and, 95–98, 167–170, 178–179, 186–187
attention-data currency, value of, 65, 193, 199–201; chart, 87; damage assessments and, 221, 235; exploitation of, 86, 157, 168, 169, 216; fair value, 177–178; "free" fallacy, 161, 178, 179–181; "meaningful social interactions" metric, 157; monopolies determine, 33–35, 129, 164, 167–169, 188, 218

Audience Network, 81, 96, 102, 194
Australia, 37, 66
Australian Consumer and Competition Commission (ACCC), 37

Bakshy, Eytan, 184
behavioral economics, 76
behavioral profiling, 59, 86, 127, 187; behavior data, protection of, 91, 201, 202, 209–210; digital social contract and, 202–203, 209–210; discrimination and, 156; engagement data and, 99–100; facial-recognition software use in, 93; psychographic traits, 18; temporally sensitive data and, 120–124; through algorithms, 132–133, 156, 188, 233; through social-graph analysis, 101; transparency and, 233. *See also* algorithmic discrimination; algorithms; data inferences
Belgium, 38
Bell Curve, The (Murray), 142
Benioff, Marc, 54–55
Berners-Lee, Tim, 28, 29
Bertrand, Marianne, 134
Best Buy, 177
Bezos, Jeff, 28–29
Biden, Joe, 130
big-data economy. *See* attention-data currency
biometric identifiers, 38, 89, 91–93
Blumenthal, Richard, 118
Bolsonaro, Jair, 9–10
Booking, 45
Bose, Subhas Chandra, 5
Bostrom, Nick, 235
Brandeis, Louis, 67, 117
Brazil, 9–10, 38, 66, 80
break-up policies, 164, 197, 214–217, 219–220, 221, 243. *See also* antitrust law and regulatory framework
brick-and-mortar retailers, 92–93

Calcutta, 4–7
California, privacy law in, 38, 76, 91, 202, 206, 207
California Notice of Security Breach Act, 89–90
Callon, Michel, 150
Cambridge Analytica incident, 17–19, 66, 91, 130–131, 229–230, 232; after effects of, 189–190. *See also* election interference; Facebook, disinformation spread on
Canada, 66
Cannon, Tyrone, 152
capitalism. *See* free market capitalism
Capitalisme et Schizophrénie (Deleuze and Guattari), 149
Cartel Office (Bundeskartellamt) suit (Germany), 179–181
CBS All Access, 176
cell-tower data, 106
Center for International Media Assistance (National Endowment for Democracy), 94
Chaslot, Guillaume, 33, 136
chief executive officers. *See* corporate executives; Zuckerberg, Mark
Children's Online Privacy Protection Act, 208
Chile, 80
China, 16, 66, 246–247; Apple and, 52, 53–54, 83, 85, 243–244; global regulations for firms based in, 242–244
Chrome browser, 98–99, 172
Cicilline, David, 36
civil disobedience, 152
Civil Rights Act (1964), 137
civil rights laws, 136–139, 156
Clayton Act, 197, 222

Clegg, Nick, 214
Clinton, Bill, 104, 105
Clinton, Hillary, 2–3, 34–35, 61
cloud-service providers, 48, 53
Code of Fair Information Practices (1973), 67–69
collective action, 236–237
colonialism and postcolonialism, 4, 5
Commerce Department, 246
commercialized decisionmaking, 126, 146–155; radical, 148–154; as rhizomatic, 149–150; transaction costs, 154–155. *See also* decisionmaking
Commission nationale de l'informatique et des libertés (CNIL), 37
Communications Decency Act (1996), 13
competition, promoting, 147–148, 186, 189–190. *See also* anticompetitive practices; antitrust law and regulatory framework; digital social contract, competition issues; monopolies
confirmation bias, 142–143
Congress, 14, 39–40, 66, 118; congressional hearings for Zuckerberg/Facebook, 1–2, 35, 36, 39, 40–41, 129–130, 243
consent framework for digital social contract, 203–204, 231–232
conservative media propaganda loop, 184
consumer attention. *See* attention-data currency
consumer-harm oriented antitrust standards, 212, 221
consumer internet, criticism of, 12–14, 30–38, 179; antitrust inquiries, 33–35; by current and former industry employees, 32–33; Facebook, 32–33, 37–38, 54; Google, 34–35, 38; new thinking on, 30–34; by organizations, 32. *See also* digital platforms, as harmful; Facebook, criticism of; Google, criticism of
consumer internet business model, pillars of. *See* advertising, targeted; algorithms; attention-data currency; data collection
consumer internet hegemony. *See* Facebook-Google-Amazon market dominance
Consumer Privacy Bill of Rights (2012), 117, 121
Consumer Privacy Bill of Rights Act (2015), 117–120, 121
contact lists, 102, 103
content dissemination network, radical commercialization of, 152–153
content policy reform, 237–242
content production and curation, 128, 146–147; Facebook feed algorithms, 58, 60, 144, 148, 187, 233; machine-learning models for, 132, 133; meaningful social interaction and, 157–159; radical commercialized decisionmaking in, 152. *See also* advertising, targeted; Facebook feed
Cook, James, 101
Cook, Tim, 50–52, 54, 83–85. *See also* Apple
cookies, 47, 94–99, 154, 204, 209
Copernicus, Nicolaus, 26
Coronavirus (COVID-19), 21–22, 43, 250
corporate executives, comments by, 27–30, 48–55; Benioff, 54–55; Clegg, 214; content policy reform and, 241–242; Cook, 50–52, 54, 83–84; on data privacy, 122; on disinformation dissemination, 15–16; "free" services claims, 161,

corporate executives (*cont.*)
179; industrial obscurantism in, 79, 82–84; on mobile ecosystem growth, 109–111; Pichai, 82, 109; Sandberg, 15, 16, 179; Smith, Brad, 55; at social media crisis talks, 14; strategic propaganda, 15–16, 48–56, 79, 82–85; Zuckerberg, 1–2, 35, 40, 79–82, 130, 147, 157, 243
correlation implies causation fallacy, 142
Costco, 177
credit card data, 107–108
Cruz, Ted, 35
cryptocurrency, 37, 185
Customs and Border Protection, DHS, 33

DailyMotion, 242
data, protection of: anonymization, 210–211; of behavioral data, 91, 201, 202; of biometric identifiers, 89; Cartel Office (Bundeskartellamt) suit and, 181; in China, 52–53, 66; Code of Fair Information Practices on, 67–69; Consumer Privacy Bill of Rights, 117–120, 121; data breaches, 89–90, 120–122, 211; historical demand for, 117–120; of PII, 89–91; through privacy engineering, 124–125. *See also* General Data Protection Regulation (GDPR, EU); regulatory protections
data, training, 131, 139–142, 143
data, value of. *See* attention-data currency
data analysis. *See* algorithms; data inferences
data breaches, 89–90, 120–122, 211. *See also* Cambridge Analytica incident
data brokers, 106, 123, 165, 189–190, 206–207
data collection, 59–60, 65–126, 154–156, 187, 220; Amazon, 165; Android, 109; Apple, 50–51, 54, 83; citizen data portability, 223; consumer ignorance of, 109, 124, 177–178; criticism of, by corporate executives, 50–52, 54–55, 83; do-not-track efforts, 98–99; electricity consumption example, 70–74; granularity levels of, 70–74, 86, 105–106, 141, 150–151; industrial obscurantism regarding, 79–85; international regulations and standards, 36–38, 52–53, 66; mergers and acquisitions affect, 222; as opt-in service, 86–88, 124, 203–204; in other industries, 43, 60; rate and quantity of, 43, 65, 120, 124, 165, 170, 178–179; sharing data, 102–103, 107–108, 112, 168, 229–230; terms of service contracts and, 85–88; through cookies, 47, 94–99, 154, 204, 209; transparency, 68, 206. *See also* algorithms; attention-data currency; Cambridge Analytica incident; Facebook, data and; Google, data and
data inferences, 47–48, 70; accuracy, 68–69, 158; by Amazon, 165; biometric data, 92; Cambridge Analytica and, 18, 91; digital social contract and, 204–205, 209–210, 233; electricity consumption example, 72–74; engagement data, 99–100, 101; location data, 105–106; smartphone data, 101, 111; social-graph data, 101–102; temporal sensitivity of, 120, 123–124; transaction data, 107;

web browsing data, 94–95. *See also* algorithmic discrimination; algorithms; behavioral profiling
data localization, 52, 66
data monetization. *See* attention-data currency; data collection
data privacy. *See* digital social contract, privacy issues; privacy, regulatory protections regarding
data sensitivity: biometric identifiers, 38, 89, 91–93; digital social contract and, 208–209; granularity levels and, 70–74, 106; PII, 89–91, 153; temporal, 120–124
data types, 89–113, 155–156; biometric identifiers, 91–93; engagement, 99–100; location, 104–106; mobile ecosystem, 108–113; PII, 89–91, 153; social-graph, 100–104; transaction, 106–108, 123, 153; web browsing and mobile use, 94–99, 172
Datta, Anupam, 144
DC Universe, 176
decisionmaking, 144, 186, 230; corporate, 17, 31, 54, 130; discriminatory, 136, 139, 150. *See also* algorithms; commercialized decisionmaking
Defense Advanced Research Projects Agency, 229
Deleuze, Gilles, 149
Dell, 45
Department of Commerce (U.S.), 153
Department of Health, Education, and Welfare (U.S.), 67–68
Department of Homeland Security (U.S.), 33
Department of Housing and Urban Development (U.S.), 156
developed nations, 245–246
developers, 32
developing nations, 41–42, 79–80, 245–246
digital advertising. *See* advertising
digital platforms, as beneficial, 8, 21, 27–32, 150
digital platforms, as harmful, 8–14, 30–39, 43–45; addiction to social media, 33, 54–55, 59, 185; international inquiries on, 36–38; scholarship on, 30–38; U.S. politicians on, 35–36. *See also* algorithmic discrimination; behavioral profiling; consumer internet, criticism of; disinformation dissemination; Facebook, criticism of; Google, criticism of; hate speech and violence
digital social contract, 200–251; advertising regulations and, 200, 202, 204, 209–210, 219, 221, 224; attention-data currency addressed through, 200–201; bipartisan support needed for, 202–203, 234–235; collective action, importance of, 236–237; content policy reform as distraction from, 237–242; GDPR as model for, 202–211; international implications, 238, 242–248, 250. *See also* regulatory protections
digital social contract, competition issues, 201, 212–223; antitrust enforcement and reform, 221–222; break-up policies to address, 214–217, 219–220; citizen data portability and, 223; consumer internet regulated as natural monopoly, 217–220; mergers and acquisitions, scrutiny over, 222; targeted reforms proposals, 219–223. *See also* antitrust law and regulatory framework; monopolies

digital social contract, privacy issues, 201–212; administrative and operational stipulations, 211; behavioral profiles and, 202–203, 209–210; consent, choice, and control in, 203–204, 231–232; data anonymization, 210–211; data sensitivity, 208–209; redress, 206–207; right to be forgotten, 207–208; technical functionality requirements, 205; transparency, 205–206. *See also* General Data Protection Regulation (GDPR, EU); privacy, regulatory protections regarding

digital social contract, transparency issues, 201, 205–206, 223–233; algorithm design, 232–233; data sharing with researchers, 229–230; ethical code for AI design, 230–231; FTC audits, 231–232; political advertising, 226–229, 233, 234, 241. *See also* transparency

disclosure. *See* transparency

discrimination: civil rights laws, 136–139, 156; discriminatory decisionmaking, 136, 139, 150; targeted ads and, 136, 138, 141, 145, 155–156. *See also* algorithmic discrimination

disinformation dissemination, 43, 78, 95, 113–115, 152, 196; network effect and, 184; Sandberg on, 15, 16; Trump and, 34–35, 129. *See also* election interference; Facebook, disinformation spread on

disparate treatment and disparate impact cases, 136, 137–139, 142, 156, 158, 160. *See also* algorithmic discrimination

do-not-track efforts, 98–99

dopamine, 59

Dorsey, Jack, 226

dot-com bust, 182

drift, problem of, 142–143, 144

Dutch Data Protection Authority (Autoriteit Persoonsgegevens), 38

Eagle, Nathan, 101

Easterbrook, Frank, 212–213

eBay, 45, 177

echo chamber effect, 79, 184

Eckles, Dean, 184

e-commerce. *See* Amazon

e-commerce ranking of top 10 companies by retail sales shares, 177

economic competition. *See* anticompetitive practices; antitrust law and regulatory framework; digital social contract, competition issues; Facebook-Google-Amazon market dominance; monopolies

economic policy, 238, 240

economic satiation, 184

economy, of Calcutta, 4–7

Edge browser, 172

Edgett, Sean, 40

education, academia and knowledge keeping, 23–26

Egypt, 21

election interference, 1–3, 34–35; congressional scrutiny, 39–40, 66; disinformation on Facebook, 2, 3, 15–16, 60–61; Russian infiltration of social media platforms, 2, 3, 60–63, 113–115. *See also* Cambridge Analytica incident

elections, in Brazil, 9–10

electricity consumption data, 70–74

electricity monopoly, 191–193, 213, 217

Electronic Communications Privacy Act (1986), 117, 208

enforcement. *See* regulatory protections, enforcement of

engagement data, 18, 99–100, 101, 138, 158
engagement maximization, 58–59, 60, 99–100, 127, 130, 194–196; Facebook, 15, 18, 81, 195; facial-recognition technology and, 93; "meaningful social interaction" and, 157; social-graph data and, 101; YouTube "rabbit hole" effect and, 10, 33, 59, 134, 136, 202
engineers, 32
Epoch Times, The, 95
Equifax, 165
"ethnic affinity" groups, 145
European Union (EU), 36–38, 66, 85, 119. *See also* General Data Protection Regulation (GDPR, EU)
Expedia, 45, 47
Explainable Artificial Intelligence program, 229

Facebook: account deactivation, 185; addiction, 185; algorithm use, 38, 58, 93, 145, 156, 195; Cartel Office (Bundeskartellamt) suit, 179–181; congressional hearings, 1–2, 35, 36, 40, 129–130; default visibility settings, 146, 195; development of, 28; echo chamber effect, 79; fines against, 113, 196; high-value audiences, 159, 160; industrial obscurantism by, 79–82; "meaningful social interactions" metric, 79, 147, 157–159; "move fast and break things" maxim, 128, 225; "oversight board" of, 242; Sandberg as COO, 15, 16, 30, 50, 179; violence and hate speech on, 9, 14, 79, 213, 214, 219, 237. *See also* Cambridge Analytica incident; Zuckerberg, Mark
Facebook, advertising and, 178–179; ACLU on, 136; Amazon as perceived challenge to, 164–165, 166–170, 178, 186; Audience Network, 81, 96, 102, 194; behavioral profiles in, 156; data brokers, 189–190; Epoch Times advertising on, 95; Facebook pixel, 96–98; political ads, 95, 130, 226, 233, 234; proxy data in, 145; revenue from, 48, 49, 56, 57, 178, 180, 186; transparency and, 233, 234; vertical integration and, 81, 82. *See also* advertising
Facebook, as social media monopoly. *See* Facebook-Google-Amazon market dominance
Facebook, criticism of: by ACLU, 136; by Apple, 51, 83–84; Cartel Office (Bundeskartellamt) suit, 179–181; by current and former employees, 32–33, 214; effects of, 189–190, 225; hate speech and violence, 9, 14, 79, 213, 214, 219, 237; international, 14, 37–38; Partner Categories shut down after, 189–190; *60 Minutes* exposé, 32; terms of service complaints, 85. *See also* Cambridge Analytica incident; consumer internet, criticism of
Facebook, data and, 46, 58, 80–81, 82, 91, 231–232; attention-data currency and, 96–98, 170–171, 178, 179; Cartel Office (Bundeskartellamt) suit, 179–181; Cook on, 54, 83–84; current and former employees criticize, 32; engagement maximization in, 195; Facebook pixel, 96–98; international criticism of, 37–38; mobile use surveillance, 111–112; Partner Categories, 189–190; privacy-program teams, 232; social-graph data, 102; terms of service contracts, 85; through feeds, 47, 60, 209; transparency,

Facebook (*cont.*)
226–227, 230, 233, 234; WhatsApp data, 102–104. *See also* Cambridge Analytica incident; digital social contract, privacy issues

Facebook, disinformation spread on, 60–63, 79, 81–82; criticism, 37, 84; election interference, 2, 3, 15–16, 60–61; proposed solutions, 41, 214, 226–227; transparency regulations and, 226–227. *See also* Cambridge Analytica incident; disinformation dissemination

Facebook, revenue and market share, 45, 46, 165, 180–181; from advertising, 48, 49, 56, 57, 178, 180, 186; attention-data currency and, 96–98, 170–171, 178, 179; profit margins, 55–57; transparency perceived to impede, 233; vertical integration and, 58, 82, 193. *See also* revenue and market share rankings

Facebook feed: curation algorithms, 58, 60, 144, 148, 187, 233; data collection through, 47, 60, 209; default settings, 146, 195; hate speech and violence in, 9, 14, 79, 213, 219, 237; "meaningful social interactions" claims, 79, 147, 157–159; subscription service suggestion, potential effects on, 41; targeted ads in, 41, 60, 79, 148, 187. *See also* algorithms, recommendation and content curation

Facebook-Google-Amazon market dominance, 47, 163–170; advertising and, 164–170, 186, 187–189; attention-data currency and, 95–98, 167–170, 178–179, 186–187; breakup policies, 214–215, 219, 243; Cartel Office (Bundeskartellamt) suit, 179–181; China and, 243–244; data privacy and, 115–124, 179; fines levied, 37, 38, 113, 196; first-mover advantage, 147–148; mobile ecosystem data and, 108–113; as natural monopolies, 191, 213, 215–219; network effect and, 182–185, 198, 218; smaller rivals, 189–190, 191; temporally sensitive data and, 120–124; terms of service contracts, 86; transaction data and, 106–108; Trump on, 34–35; as two-sided platforms, 168, 216; value propositions in, 147; vertical integration and, 193. *See also* anticompetitive practices; antitrust law and regulatory framework; digital social contract, competition issues; monopolies

Facebook Live, 14

Facebook Messenger, 58

Facebook subsidiaries and service offerings, 13; Audience Network, 81, 96, 102, 194; cross-platform texting, 19–20; in developing nations, 79–80; Facebook Live, 14; Libra cryptocurrency, 37, 185; Messenger, 58, 173, 187, 216; mobile ecosystem data and, 111–112; Oculus, 64, 185; Onavo, 111; Partner Categories, 189–190; Workplace, 57. *See also* Facebook feed; Instagram; WhatsApp

facial-recognition technology, 38, 92–93

Fair Housing Act, 156

Fair Information Practice Principles (Federal Trade Commission), 69

fairness/justice arguments, 134–136

fake news. *See* disinformation dissemination

Family Education Rights and Privacy Act, 208
Farivar, Cyrus, 112
Federal Cartel Office of Germany (Bundeskartellamt), 179–181
Federal Trade Commission (FTC), 35, 112–113, 153, 215; Consumer Privacy Bill of Rights, 118, 120; Federal Trade Commission Act, 197, 213; industry auditing by, 231–232; regulation enforcement by, 69, 118, 119–120, 196, 221. *See also* regulatory protections
Feinstein, Dianne, 61
fines levied against digital platforms, 37, 38, 113, 196
Firefox, 172
fourth estate. *See* journalism and news reporting
France, 37
Franken, Al, 118
freedom of information, 207
freedom of speech, 130, 207, 238. *See also* hate speech and violence
"free" fallacy, 161, 178, 179–181
free market capitalism, 70, 116, 128–129, 135–136, 161, 200; radical commercialized decisionmaking and, 148–154. *See also* digital social contract; Facebook-Google-Amazon market dominance; monopolies
FTC. *See* Federal Trade Commission (FTC)
Furman, Jason, 213

Galpaya, Helani, 80
General Data Protection Regulation (GDPR, EU), 36–37, 66, 85; consent framework, 204; data anonymization and, 210–211; data inferences and, 210; as digital social contract model, 202–211; right to be forgotten, 207; right to rectification, 207; sensitive data protections, 208–209. *See also* regulatory protections
Germany, 37–38, 179–181
Ghonim, Wael, 21
global connectivity: Calcutta and, 4–7; regulatory issues, 238, 242–248, 250
Global Positioning System (GPS), 104
Google, 2, 14, 132, 147, 215; Android OS, 108–111; China and, 244; development of, 28; election interference through, 40, 60–61; facial-recognition technology, 93; industrial obscurantism by, 82–83; network effect, 183; subsidiaries and service offerings, 48, 57, 103, 108–111, 135, 174. *See also* Facebook-Google-Amazon market dominance; YouTube
Google, advertising and, 107–108, 154, 178–179; Amazon as perceived challenge to, 164–165, 166–170, 178, 186; revenue from, 48, 49, 165, 178, 180, 186, 244; tracking-and-targeting hegemony, 98–99
Google, as search engine monopoly. *See* Facebook-Google-Amazon market dominance
Google, criticism of: by Apple, 50, 54; by current and former employees, 30–31, 33; international, 37, 38; terms of service complaints, 85; by Trump, 34–35
Google, data and, 43; criticism of, 50, 54; international criticism of, 38; mobile use surveillance, 112; regulatory protections, 82–83; transaction data, 107–108. *See also* digital social contract, privacy issues

Google, revenue and market share, 45, 46; from advertising, 48, 49, 165, 178, 180, 186, 244; profit margins, 55–57; vertical integration, 193. *See also* revenue and market share rankings
Google Cloud, 48
Google Earth, 174
Google Maps, 174
Google Photos, 135
Google Play, 103
Google subsidiaries and service offerings, 48, 57, 103, 108–111, 135, 174
government reforms. *See* regulatory protections
Graham, Lindsey, 35, 202, 212
Gramm-Leach-Bliley Act, 208
Groupon, 45
Guattari, Félix, 149

Harris, Tristan, 30–31, 59
Hart-Scott-Rodino Antitrust Improvements Act, 222
Harvard Business School, 55
Hatch, Orrin, 39, 40
hate speech and violence, 8–9, 38, 43, 238; Facebook and, 9, 14, 79, 213, 214, 219, 237; vigilante violence, 11, 44; on YouTube, 14, 59
Hawley, Josh, 202
Health Insurance Portability and Accountability Act, 208
Hearst Corporation, 146–147
Hewlett-Packard, 45
high-value audiences, 159–160
The Home Depot, 177
homophily principle, 101–102
Honest Ads Act proposal, 36, 234, 241
Hosanagar, Kartik, 99–100
House Judiciary's Subcommittee on Antitrust, Commercial, and Administrative Law, 35, 36
Hughes, Chris, 214
Hulu, 175, 176
human rights standards, 52, 83
hyperbolic discounting theory, 76–78, 88, 118

IBM, 18, 46
identity data, 89–91, 153. *See also* data collection
identity theft, 90
Illinois, privacy law in, 38, 93
Illinois Biometric Information Privacy Act, 93
India, 11, 66, 80; Calcutta, 4–7
Indonesia, 80
industrial obscurantism, 79–85, 88, 118, 205–206
influence market, 16–17
information asymmetry, 88, 118
information gathering. *See* data collection
innovation, copying, 51–52, 112, 129, 194, 197, 218–219. *See also* anticompetitive practices
Instagram: advertising revenue, 49; age demographics of users, 160; attention-data currency and, 167, 168, 171; cross-platform texting on, 19; data and, 77–78, 158, 167, 168, 171; facial-recognition technology, 93; mergers and acquisitions and, 214, 222; network effect and, 58, 187. *See also* Facebook
instant gratification, 77
Intel, 45
intellectual obscurantism, 79
International Grand Committee on Disinformation and Fake News, 37
International Telecommunications Union (UN), 246
internet barons. *See* corporate executives, comments by; Facebook-

Google-Amazon market dominance
internet browsers, ranked by market share, 172
Internet Corporation for Assigned Names and Numbers, 246
Internet Explorer, 172
Internet.org, 79–80
investments, 57, 84
iOS, 108, 110
iPhones, location data and, 104
Islam, 14
Italy, 38

Japan, 37, 66
journalism and news reporting, 12–14; centralized power, diminishment of, 146–147; ethics codes, 151; Facebook pixel use on sites, 95, 98
Justice Department (U.S.), 35, 113; Antitrust Division, 215, 221–222

Kahneman, Daniel, 135
Kimmelman, Gene, 40
Kituyi, Mukhisa, 37
Kizilcec, René, 184
Knetsch, Jack, 135
knowledge, academia as keepers of, 23–26
Kogan, Aleksandr, 229–230
Kuhn, Thomas, 23–26

Lanier, Jared, 41
Latin America, 66
Latour, Bruno, 150
Law, John, 150
law and legal justice: antidiscrimination laws, U.S., 137, 143, 144, 145; fairness equated with justice, 135–136. *See also* antitrust law and regulatory framework; regulatory protections, enforcement of
Lazer, David, 101
learning models. *See* algorithms
Lee, Dokyun, 99–100
legislation. *See* regulatory protections
Lessig, Lawrence, 168
Libra cryptocurrency, 37, 185
Lindy effect, 189
LinkedIn, 49, 171
Lisovich, Mikhail, 72–74
LiveLeak, 14
location data, 104–106
Logic of Collection Action, The (Olson), 236
Lowe's Home Improvement, 92–93, 107
Lynn, Barry, 31, 34, 212

machine bias. *See* algorithmic discrimination
machine-learning models. *See* algorithms
Mactaggart, Alastair, 202
Macy's, 177
mapping apps, ranked by active users, 174
Mapquest, 174
market decisions, 107, 108
marketing. *See* advertising
market share. *See* revenue and market share rankings
Markey, Ed, 118
Marriott/Starwood data breach, 122
Marx, Karl, 161
MasterCard, 107–108
McCain, John, 3636
McNamee, Roger, 30, 41
McPherson, Miller, 101
"meaningful social interaction" claims, 79, 147, 157–159
media, past/other forms of, 60, 149–150; print media, 12, 13, 146–147; regulation of, 151

mergers and acquisitions, 214, 219, 222
Messenger (Facebook), 58, 173, 187, 216
Metcalfe's law, 183
Microsoft, 14, 57, 93, 180; privacy regulation advocacy by, 46, 55; revenue of, 49, 56
Microsoft Azure, 48
Mill, John Stuart, 8
Miller, Carl, 206
mobile ecosystem data, 108–113
mobile messenger apps, ranked by active users, 173
mobile operating systems, market shares, 110
Mohan, Neal, 14
monopolies, 190–196; attention-data currency and, 33–35, 129, 164, 167–169, 188, 218; effects of, 168, 193–194; legal definition, 197–198; monopoly rents, 86, 168, 200; natural, 190–191, 213, 217–220; network effect in creation of, 183–184, 187, 191–192, 198, 218; telecommunications, railroad, and electrical industries, 119, 161, 182, 191–192, 213, 217–218; vertical integration and, 192–193. *See also* anticompetitive practices; antitrust law and regulatory framework; digital social contract, competition issues; Facebook-Google-Amazon market dominance
Mullainathan, Sendhil, 134
Mundt, Andreas, 181
Murray, Charles, 142
Myanmar, 8–9, 237

Nair, Harikesh, 99–100
Narayanan, Arvind, 100
National Endowment for Democracy, 94
National Institute of Standard and Technology's Cybersecurity Framework, 231
natural monopolies, 190–191, 213, 215–220. *See also* antitrust law and regulatory framework; digital social contract; Facebook-Google-Amazon market dominance; monopolies
navigation technology, 104, 174
Netflix, 45, 128, 175, 176
the Netherlands, 38
net neutrality, 80
network effect, 58, 182–185, 187, 191–192, 198, 218
New America Foundation, 31
news. *See* journalism and news reporting
New York Times, 12, 31, 55, 82, 95, 98
New Zealand, 14
Nickerson, Raymond, 143
Nielsen, 18, 123
Nigeria, 80
Nissenbaum, Helen, 153

Obama, Barack, 117–118
Obama administration, 66, 117–120, 121, 234, 246
Oculus, 64, 185
Olson, Mancur, 236, 237
Onavo (Israeli web analytics firm), 111
on-platform engagement data, 99–100. *See also* engagement maximization
Open Markets Institute, 31, 34, 36, 190
Opera, 172
operating systems, 108–111
Oracle, 46, 190
Organization for Economic Cooperation and Development, 194

Pai, Ajit, 234
Palihapitiya, Chamath, 32–33
Parakilas, Sandy, 32
Partner Categories (FB data broker), 189–190
Pennycook, Gordon, 152
Pentland, Alex, 101
personal data. *See* data collection
personally identifiable information (PII), 89–91, 153; biometric data, compared, 91–92
photos, facial-recognition technology and, 92–93
Pichai, Sundar, 50, 82–83, 109
Pinterest, 171
pixel (Facebook), 96–98
Podesta, John, 2–3, 117
policy. *See* regulatory protections
political advertising. *See* advertising, political
political elections, Brazil, 9–10
political elections, U.S. *See* election interference
politics and partisanship: anticonservative bias, 33, 34–35, 44; bipartisan criticism of consumer internet, 36; bipartisan regulatory support, need for, 202–203, 234–235; conservative media, propaganda feedback loop, 184; freedom of speech and, 238; Obama administration, 66, 117–120, 121, 234, 246; political polarization, 6, 130, 234–235, 241; Trump administration, 33, 66, 113
populism and radicalization, 9–11, 43–44
Prelec, Drazen, 78
presidential election (2016). *See* election interference
PricewaterhouseCoopers, 232
print media, 12, 13, 146–147
privacy, definition of terms, 67, 81–82
privacy, regulatory protections regarding, 17, 74–91, 115–120; Apple and, 50–52; behavioral data, 91, 201, 202, 209–210; biometrics, 93; California, 38, 76, 91, 202, 206, 207; Code of Fair Information Practices, 67–69; Consumer Privacy Bill of Rights, 117–120, 121; criticism of, 120, 153; debates, 65–70; *vs.* engineered solutions, 124–125; failure to pass in U.S., 76, 118–120, 122, 128–129; history of demands for, 32, 117; hyperbolic discounting theory, privacy paradox and, 76–78, 88, 118; industrial obscurantism regarding, 79–85; Microsoft and, 46, 55; PII data *vs.* other types, compared, 89–91; terms of service contracts and, 85–88. *See also* digital social contract, privacy issues; regulatory protections
privacy engineering, 124–125
privacy paradox, hyperbolic discounting theory and, 76–78
"privacy prices," 177–178
privacy-program teams, 232
profit margins, 55–57, 84, 161. *See also* revenue and market share rankings
ProPublica, 145
proxy generation in algorithms, 144–145
psychographic traits, collection of, 18. *See also* behavioral profiling
public utilities regulations, 119, 161, 182, 191–192, 213, 217–218
publish-or-perish culture in academia, 25
purpose specification, of data collected, 68
Putin, Vladimir, 2, 61, 113–114

QQ Mobile, 173
Quarate Retail Group, 177
Quora, 171

racism, algorithms and, 134, 135, 144, 145
radical commercialized decisionmaking, 148–154
radicalism, 9–11, 43–44
railroad monopoly, 191–192, 213, 217
Ramo, Joshua Cooper, 7
Rand, David, 152
Rawls, John, 134–135
recommendation algorithms. *See* algorithms, recommendation and content curation
reddit, 171
redress (correction of inaccurate data), 68–69, 206–207
Reed's law, 183–184
regulatory protections: against algorithmic discrimination, 161, 208, 223–227, 229–230, 232–233; bipartisan support needed for, 202–203, 234–235; cryptocurrency, 185; news media, 14; telecommunications, railroad, and electrical monopolies, 119, 161, 182, 191–192, 213, 217–218. *See also* advertising, regulatory protections regarding; data, protection of; digital social contract; Federal Trade Commission (FTC); General Data Protection Regulation (GDPR, EU); privacy, regulatory protections regarding
regulatory protections, anticompetitive practices and. *See* antitrust law and regulatory framework; digital social contract, competition issues
regulatory protections, enforcement of, 236, 250; antitrust law, 212–213, 221–222; fines, 37, 38, 113, 196; by FTC, 69, 118, 119–120, 196, 221; in Illinois, 38; international, 36–38, 52–53, 66, 119, 206; lack of, 69, 82–83, 112, 113, 116–119; regression of, 163
retail online sales, ranked by sales share, 177
revenue and market share rankings, 40–41, 45–57, 59, 187, 189; antitrust law and, 215, 216; Apple, 45–46, 50–55, 56, 128, 177; Chinese market and, 244–245; corporate executives' comments and, 48–51, 54–55; global mobile messenger apps, 173; inclusion criteria, 45–47, 54; internet browsers, 172; mapping apps, 174; mobile OS use, global, 110; online video viewing, 175; original video content demand, 176; in print news media, 12, 13; retail online sales, 177; social media platforms, by visits, 171; Twitter, 45, 47, 171. *See also* advertising revenue; Amazon, revenue and market share; Facebook, revenue and market share; Google, revenue and market share
right to be forgotten, 207–208
right to privacy. *See* privacy
Rohingya genocide, 8–9, 237
Roose, Kevin, 112
Rosseau, Jean-Jacques, 200
Roy, Deb, 152
Russia, 245–246, 246–247
Russian infiltration of social media platforms, 2, 3, 60–63, 113–115. *See also* election interference

Safari, 172
Salesforce, 45, 54–55, 57
Salgado, Richard, 40

sanctions policy, 247
Sandberg, Sheryl, 15, 16, 30, 50, 179
SAP, 46
Schatz, Brian, 39
Schiebinger, Londa, 134
Schmidt, Eric, 28, 29
Schumpeter, Joseph, 12
science and engineering, fields of, 23–24, 25
Scott, Ben, 177
selection bias, 140–141
sensitive data. *See* data sensitivity
sentiment analysis, 124
Sepulveda, Danny, 246
service offerings, 59, 187, 205; Amazon, 165; Google, 48, 57, 103, 108–111, 135, 174. *See also* Facebook subsidiaries and service offerings
Sessions, Jeff, 35
Shapiro, Carl, 212–213
Sherman Act, 197, 221
Singapore, 37–38
60 Minutes (television show), Facebook exposé on, 32
smart meters, 71
smartphones, 101; location data, 104–106; mobile ecosystem data, 108–113
smart TVs, 100
Smith, Brad, 50, 55
Smith, Chris, 109–111
Smith-Lovin, Lynn, 101
Snap, advertising and, 166, 189, 190
Snapchat, 173, 180, 189
Snowden, Edward, 202
social-graph data, 100–104
social media addiction, 33, 54–55, 59, 185
social media crisis talks, 14
social media platforms. *See* digital platforms; Facebook; Twitter
social media platforms, ranked by market share of visits, 171
Social Science One project, 230
social security numbers, 89–90
socioeconomic disparity, subscription services and, 41–42
Solon, Olivia, 112
Solove, Daniel, 76, 118
South Korea, 66
Starbucks, 93
"Statement of Enforcement Principles Regarding 'Unfair Methods of Competition'" (FTC), 221
StreetBump app, 141
Stretch, Colin, 40
Structure of Scientific Revolutions, The (Kuhn), 23–26
subscription services, 41–43
Subsecretaría de Telecomunicaciones (Chile), 80
Sunstein, Cass, 152
surveillance capitalism, 30. *See also* data collection

Tagore, Rabindranath, 5
Tarrant, Brenton, 14
Telecommunications Act (1996), 182
telecommunications monopoly, 119, 161, 182, 191–192, 213, 217–218
Telecom Regulatory Authority (India), 80
Telegram, 173
terms of service contracts, 85–88
terrorism, 14, 43–44
texting, cross-platform, 19–20
Thaler, Richard, 135
TikTok, 242
traditional news media. *See* journalism and news reporting; media, past/other forms of
training data, 131, 139–142, 143. *See also* algorithmic discrimination

transaction cost of information exchange, 154–155
transaction data, 106–108, 123, 153
transparency, 201, 205–206, 223–226; algorithms and, 232–233; Code of Fair Information Practices on, 68; Facebook and, 226–227, 230, 233, 234; in political advertising, 226–229, 233, 234, 241. *See also* digital social contract, transparency issues
Trump, Donald, 1, 2, 34–35, 61, 129
Trump administration, 33, 66, 113
Tufecki, Zeynep, 31
Twitter, 147; advertising and, 156, 166, 180, 226; election interference through, 2, 40, 60–61; engagement data, 99; market share of visits, 171; political advertising blocked on, 226; revenue of, 45, 47; social media crisis talks, 14; subscription suggestion, 42; Trump tweets, 34–35; Twitter feed, 99, 144, 152

Uber, 45, 47
United Kingdom, 37, 66
United Nations Conference on Trade and Development, 37

value propositions, 147, 157
Verifications.io data breach, 122
Verizon, 49, 166
vertical integration, 58, 81, 82, 149, 192–193, 222
Verveer, Phil, 40, 164, 246
video-streaming services: demand for original content, ranked by market share, 176; network effect, 183; online video viewing, ranked by market share, 175; terrorist acts viewed over, 14. *See also* YouTube
vigilante violence, 11, 44
violence. *See* hate speech and violence
virtual reality, 185
Vosoughi, Soroush, 152

Wachter, Sandra, 161
Walmart, 177
Warden, Pete, 104
Warner, Mark, 36, 202
Warren, Elizabeth, 36, 202, 212, 214
Warren, Samuel, 67
Washington Post, 12, 33
Wayfair, 177
Waze, 174, 222
web browsing and mobile use data, 94–99; market share of top browsers, 172; pixel code, 96–98
web cookies, 47, 94–99, 154, 204, 209
Weber, Steven, 168
WeChat, 173
Weigend, Andreas, 122
WhatsApp, 19, 39, 58, 173, 214, 216; mergers and acquisitions and, 222; network effect and, 183; social-graph data, 102–104; vigilante violence and, 11. *See also* Facebook
Wheeler, Tom, 40, 230
whistle-blowing, 32
Whitehouse, Sheldon, 118
Whole Foods, 222
Wicker, Stephen, 72–74, 105
Wilson, Harold, 33–34
Wolff, Jonathan, 135
women, hiring discrimination at Amazon, 143
Workday, 45
Workplace service (Facebook), 57
Wu, Tim, 31, 212
Wyden, Ron, 36

Xiao Han, 102

Yahoo data breaches, 122
Yahoo! Maps, 174
Yelp, 171, 180
Yoffie, David, 55
YouTube, 10, 14; advertising revenue, 49, 59; engagement data and, 100; market share of visits, 171; network effect, 183; online video viewing market share ranking, 175. *See also* Google
YouTube recommendation algorithm, 132–134, 144; discrimination and, 136; "rabbit hole" effect, 10, 33, 59, 134, 136, 202. *See also* algorithms, recommendation and content curation

Zakrzewski, Cat, 33
zero-rating programs, 79–80
Zimmer, Bob, 37
Zou, James, 134
Zuboff, Shoshana, 30
Zuckerberg, Mark, 14, 28, 50, 112, 239–241; calls for indictment of, 36, 37; congressional hearings, 1–2, 35, 39, 40–41, 129–130, 243; industrial obscurantism by, 79–82; "meaningful social interaction" claims, 79, 147, 157. *See also* corporate executives, comments by; Facebook

About the Author

DIPAYAN GHOSH is Director of the Digital Platforms & Democracy Project and Lecturer on Law at Harvard University, where he leads an interdisciplinary research program focused on examining the digital economy and its impact on society. A computer scientist by training, Ghosh previously worked on global privacy and public policy at Facebook. Prior to that, he was a technology and economic policy advisor in the Obama White House, focusing on consumer privacy and the digital economy. He has served as a fellow at the Center for Democracy and Technology and New America, both Washington, D.C.-based public policy think tanks. Ghosh received a Ph.D. in electrical engineering at Cornell University, an M.B.A. from MIT, and a B.S.E. from the University of Connecticut.